International Conference on Scanning Probe Microscopy (SPM-2017)

IOP Conference Series: Materials Science and Engineering
Volume 256

Ekaterinburg, Russia
27 – 30 August 2017

ISBN: 978-1-5108-5033-0
ISSN: 1757-8981

Printed from e-media with permission by:

Curran Associates, Inc.
57 Morehouse Lane
Red Hook, NY 12571

CURRAN ASSOCIATES INC. proceedings.com

Some format issues inherent in the e-media version may also appear in this print version.

Printed by Curran Associates, Inc. (2017)

For permission requests, please contact the Institute of Physics
at the address below.

Institute of Physics
Dirac House, Temple Back
Bristol BS1 6BE UK

Phone: 44 1 17 929 7481
Fax: 44 1 17 920 0979

techtracking@iop.org

Additional copies of this publication are available from:

Curran Associates, Inc.
57 Morehouse Lane
Red Hook, NY 12571 USA
Phone: 845-758-0400
Fax: 845-758-2633
Email: curran@proceedings.com
Web: www.proceedings.com

Table of contents

Volume 256

Scanning Probe Microscopy 2017 (SPM-2017)
27–30 August 2017, Ekaterinburg, Russian Federation

Accepted papers received: 8 October 2017
Published online: 20 October 2017

Preface

New Methods of Scanning Probe Microscopy

Scanning Probe Microscopy in Biology and Medicine

Scanning Probe Microscopy in Materials Science

SPM Analytical Methods and Probe Lithography

Ferroics

IOP Conf. Series: Materials Science and Engineering **256** (2017) 011001 doi:10.1088/1757-899X/256/1/011001

Guest Editorial

These Proceedings contain a collection of papers based on the oral and poster presentations at the **International Conference "Scanning probe microscopy" (SPM-2017)**, which was held in Ekaterinburg, Russia, August 27-30, 2017.

International Conference SPM-2017 covered a wide range of scientific topics:

- New Methods of Scanning Probe Microscopy,
- Scanning Probe Microscopy in Materials Science,
- Scanning Probe Microscopy in Biology and Medicine,
- Scanning Probe Microscopy Analytical Methods and Probe Lithography,
- Scanning Probe Microscopy Data Processing and Piezoresponse Force Microscopy,
- Ferroics.

The conference was organized by the School of Natural Sciences and Mathematics and Ural Center for Shared Use "Modern Nanotechnology" of the Ural Federal University.

The objective of SPM-2017 was to provide an international and interdisciplinary exchange of information on a wide range of topics related to modern state-of-the-art and future development of various Scanning Probe Microscopy techniques in different fields of science in order to ensure effective cooperation among scientists and specialists from universities, research institutions and industry.

More than 150 participants from 8 countries: Belorussia, China, Germany, Ireland, Israel, Kazakhstan, Portugal, and Russia presented their results. The conference comprised 162 presentations, including 14 invited, 44 oral and 104 posters.

The organizers are very grateful to the conference sponsors and exhibitors: Taylor & Francis Group, Russian Foundation for Basic Research, NT-MDT Spectrum Instruments, Melytec, Promenergolab, Ostec ArtTool, OPTEC, INTERTECH Corporation, ConeTech, SITEC, IMC Systems, and Alexander Lukanin.

Vladimir Shur
Victor Mironov
Chairmen of **SPM-2017**

IOP Conf. Series: Materials Science and Engineering **256** (2017) 011001 doi:10.1088/1757-899X/256/1/011001

SPM-2017 Committees

Conference Chairmen

Prof. Vladimir Shur
Prof. Victor Mironov

Local Organizing Committee

Prof. Vladimir Shur
Mrs. Alevtina Shur
Mrs. Elena Pelegova
Dr. Ekaterina Shishkina
Dr. Dmitry Pelegov
Ms. Victoria Pryakhina
Ms. Olga Dyachuk

Program Committee

Prof. A.A. Bukharaev, Kazan E. K. Zavoisky Physical -Technical Institute RAS, Russia
Prof. V.A. Bykov, NT-MDT Spectral Instruments, Russia
Prof. A.V. Latyshev, Rzhanov Institute of Semiconductor Physics SB RAS, Russia
Prof. V.L. Mironov, Institute for Physics of Microstructures RAS, Russia
Prof. G.M. Mikhailov, Inst. of Microelectronics Technology and High Purity Materials RAS, Russia
Prof. A.A. Saranin, Institute of Automation and Control Processes FEB RAS, Russia
Prof. V.Ya. Shur, Ural Federal University, Russia
Prof. I.V. Yaminsky, Moscow State University, Russia
Prof. S.A. Chizhik, A.V. Luikov Heat and Mass Transfer Institute NASB, Belarus
Prof. A.L. Gruverman, University of Nebraska-Lincoln, USA
Dr. A.L. Kholkin, University of Aveiro, Portugal
Prof. O.V. Kolosov, University of Lancaster, UK
Dr. V.V. Shvartsman, University of Duisburg-Essen, Germany
Dr. A.P. Volodin, University of Leuven, Belgium

IOP Conf. Series: Materials Science and Engineering **256** (2017) 011001 doi:10.1088/1757-899X/256/1/011001

List of SPM-2017 Participants

Belarus
Astashonok Andrei
Kuznetsova Tatyana
Starodubtseva Maria

China
Fu Xiaotian
Man Zhenyong
Wang Jianlong
Wei Xiaoyong
Zeng Jiangtao
Zhang Yanmin
Zhang Yong
Zhang Youjie
Zhao Kunyu
Zhao Weigang
Zhao Ye
Zhuang Yongyong

Germany
Schultheiß Jan

Ireland
Stapleton Aimee

Israel
Lubomirsky Igor

Kazakhstan
Zhakina Alma

Portugal
Kholkin Andrei

Russia
Abramov Alexander
Akbaeva Galina
Akhmatkhanov Andrei
Akovantseva Anastasia
Alekseev Prokhor
Alikin Denis
Andreeva Natalia
Anisimov Andrei
Ankudinov Alexander
Avilov Vadim
Bedin Sergei
Bukharaev Anastas
Bunin Mikhail
Butiaikin Sergey
Bykov Andrei
Bykov Viktor
Bystrov Vladimir
Cheburkin Viktor
Chezganov Dmitry
Chuvakova Maria
Dementieva Maria
Dunaevskiy Mikhail
Dyachuk Olga
Esin Alexander
Fakhrullin Rawil
Fedorovykh Vyacheslav
Filatov Dmitry
Filatova Natalia
Fokin Denis
Fomin Lev
Frolova Anastasia
Gerasimov Roman
Germanenko Alexander
Gimadeeva Lubov
Golubev Evgeny
Greshnyakov Evgeny
Gruzdev Dmitry
Gubarev Timofei
Gunina Ekaterina
Hu Qingyuan
Ilina Marina
Jityaev Igor
Karpov Vyacheslav
Kaysheva Anna
Khalisov Maxim
Kholkin Andrei
Klueva Yulia
Kolomiytsev Alexei
Kosobokov Mikhail
Kovaleva Daria
Kozlowski Alexander
Krasnoborodko Sergey
Kunkel Tatiana
Kuprina Yulia
Kuznetsov Dmitry
Linker Eduard
Lobov Ivan
Luchkin Sergey
Lukanin Alexander
Makaev Andrei
Makhmutov Artur
Mandel Arkady
Marinchuk Artem
Mikhailov Gennady
Mikhaylov Alexander
Mikhaylova Tatiana
Mingaliev Evgeny
Minin Maxim
Mironov Viktor
Morozova Anna
Nebogatikov Maxim
Neudachina Vera
Nuraeva Alla
Nurgazizov Niyaz
Nosov Alexander
Osipov Valentin
Pamyatnykh Lidia
Panova Marina
Patrakov Evgeny
Pelegov Dmitry
Pelegova Elena
Petrova Tatiana
Polubavkina Yulia
Polyakov Vyacheslav
Ponomarev Sergey
Pribora Vasiliy
Pryakhina Victoria
Pukhova Valentina
Putilov Alexei
Raevskaya Svetlana
Raevski Igor
Rassadin Alexander
Rogov Alexei
Rumyantsev Evgeny
Safiullina Aisylu
Saranin Alexander
Sazanova Tatiana
Scherbakova Olga
Sharipov Talgat
Sharov Vladislav
Shishkina Ekaterina
Shubin Andrei
Shur Alevtina
Shur Vladimir
Shuravin Andrei
Sitalo Evgeny
Skrylyov Alexander
Slautin Boris
Sokolov Denis
Stetsyura Svetlana
Sun Lina
Telny Yaroslav
Timofeeva Victoria
Tominov Roman
Turygin Anton
Udalov Artur
Useinov Alexei
Ushakov Andrei
Vasilev Semen
Vasileva Daria
Vaskina Ekaterina
Vlasov Evgeny
Vysokikh Yuriy
Yaminsky Igor
Yuzhakov Vladimir
Zagorskiy Dmitry
Zelenovskiy Pavel
Zhuravleva Yulia
Zorina Maria
Zubarev Ilya

IOP Conf. Series: Materials Science and Engineering **256** (2017) 011002 doi:10.1088/1757-899X/256/1/011002

Peer review statement

All papers published in this volume of *IOP Conference Series: Materials Science and Engineering* have been peer reviewed through processes administered by the proceedings Editors. Reviews were conducted by expert referees to the professional and scientific standards expected of a proceedings journal published by IOP Publishing.

IOP Conf. Series: Materials Science and Engineering **256** (2017) 012001 doi:10.1088/1757-899X/256/1/012001

Scanning probe nanolithography of resistive memory element based on titanium oxide memristor structures

V I Avilov*, N V Polupanov, R V Tominov, V A Smirnov and O A Ageev

Southern Federal University, Institute of Nanotechnologies, Electronics and Equipment Engineering, Research and Education Center "Nanotechnologies" 347922, Taganrog, Russia

*AvilovVI@sfedu.ru

Abstract. The paper presents the results of the formation of the resistive memory element prototype based on the memristor structures of titanium oxide using probe nanolithography by the local anodic oxidation. Such structures exhibit a memristor effect without carrying out an additional electroforming operation.

1. Introduction

Modern computing evolution is closely associated with the development of the newest fast and energy efficient nanoelectronics devices, such as the resistive random-access memory (RRAM) based on memristor structures [1-2]. Although the memristive behavior have been observed in many materials, the memristor structures based on titanium oxide exhibit high switching speed and the resistance ratio in high resistance state and low resistance state [3-4]. The formation of memristor structures for RRAM devices related with research and development of high-resolution nanolithography processes [5]. The analysis of the existing lithographic methods, such as electron-beam lithography, X-ray lithography, and focused ion beam lithography, has reached a limit. The actual solution of this problem consists in the new techniques of profiling substrate surface based on scanning probe nanolithography by local anodic oxidation in atomic force microscopy (AFM) mode [6-13]. Local anodic oxidation is one of the promising methods of fabricating oxide nanoscale structures on the surface of various conducting and semiconductor materials. It can be used for development of the nanoelectronics devices and micro- and nanosystems engineering. The aim of this work is to study memristive behavior of oxide nanoscale structures obtained by local anodic oxidation and the formation of the resistive memory element prototype based on the obtained structures.

2. Experimental

Experimental samples of titanium oxide nanoscale structures were produced (Fig. 1). First, 20 nm thick titanium film was deposited on the structure of Si/SiO_2 by magnetron sputtering. Then local anodic oxidation of titanium film was carried out using probe nanolaboratory Ntegra ("NT-MDT", Russia) with NSG 10/Pt cantilevers. As a result, titanium oxide nanoscale structures with different heights were formed. Electrical characteristics of the obtained structures were measured in the AFM spectroscopic mode, titanium film serving as a bottom electrode and AFM probe – as an upper one.

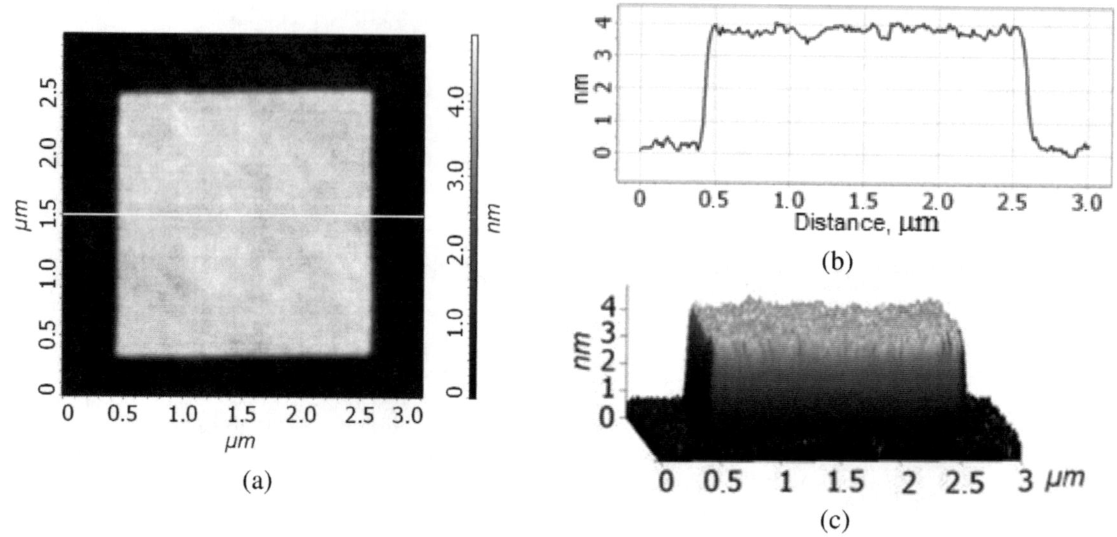

Figure 1. AFM-images of titanium oxide nanoscale structures obtained by local anodic oxidation: (a) topography, (b) profilogram, (c) three-dimensional image of the surface.

3. Results and discussion

Figure 2 presents I-V characteristics of the obtained oxide nanoscale structures. The analysis showed that the structures with height from 1.7±0.1 to 3.4±0.3 nm demonstrated memristive behavior under applied voltage pulses ±2.4 V, without electroforming process; while the structure with thickness of 3.6±0.3 nm in this voltage range did not demonstrate memristive behavior.

Moreover, it was revealed that increasing the height of the oxide from 1.7±0.1 to 3.4±0.3 nm, led to current decrease from 1.92 nA to 0.8 pA, and increase of the high resistance state value from 15 to 88 GΩ and the low resistance state value from 0.5 to 2.4 MΩ.

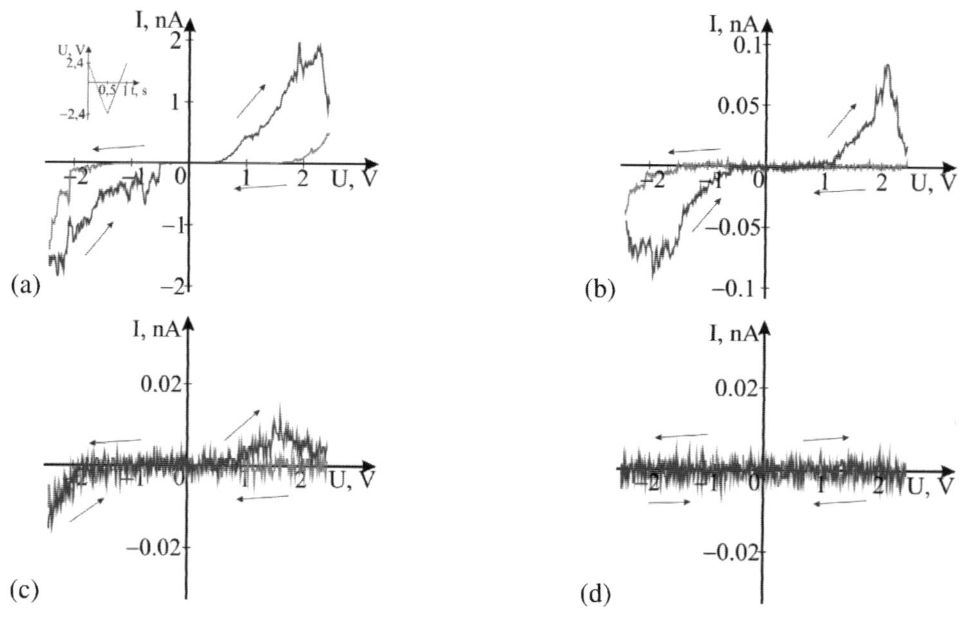

Figure 2. I-V characteristics of the oxide nanoscale structures surface with different heights: (a) 1.7±0.1 nm; (b) 2.5±0.2 nm; (c) 3.4±0.3 nm; (d) 3,6±0.3 nm.

IOP Conf. Series: Materials Science and Engineering **256** (2017) 012001 doi:10.1088/1757-899X/256/1/012001

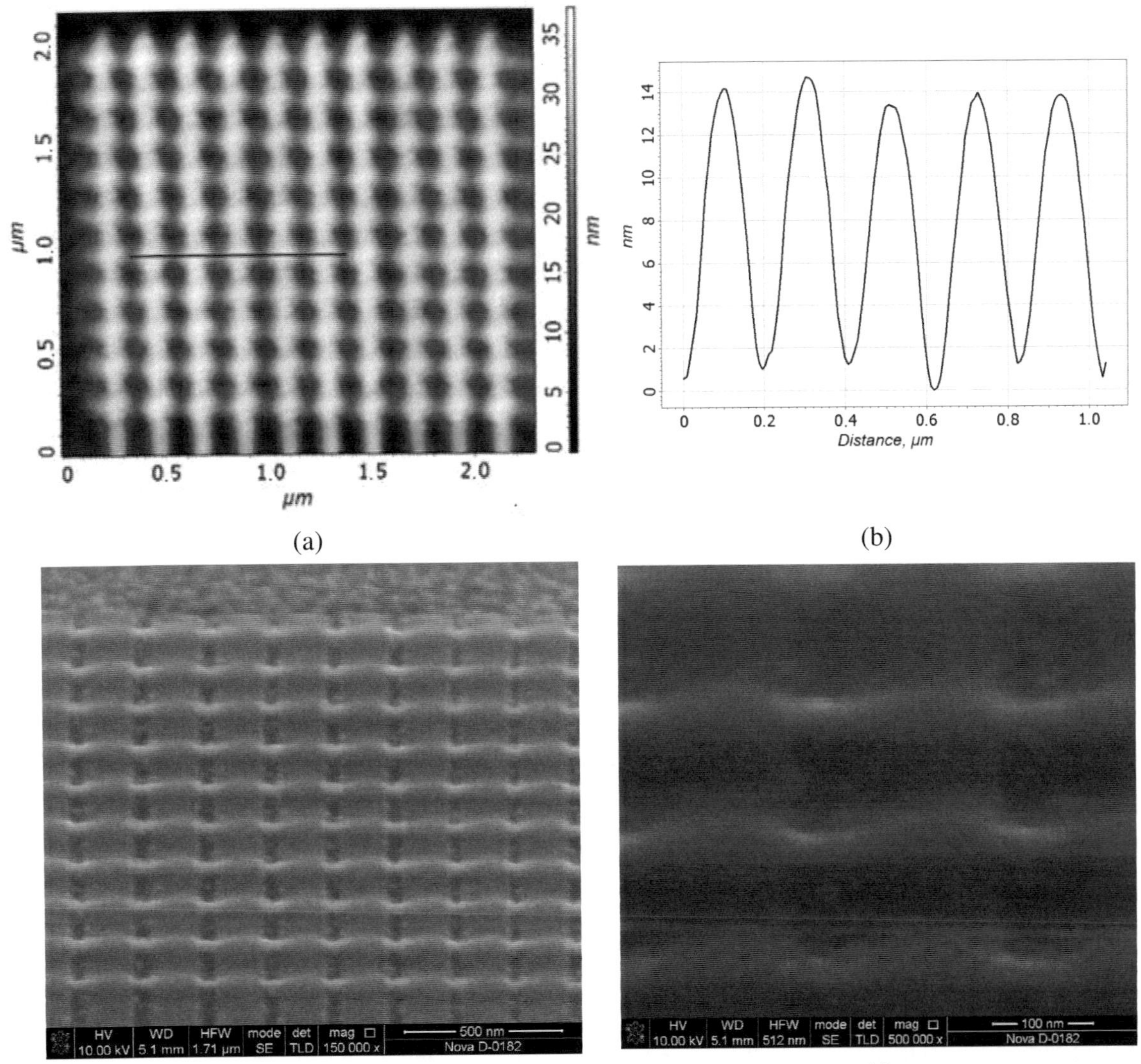

(c) (d)

Figure 3. Resistive memory element prototype based on titanium oxide memristor structures: (a) AFM-image, (b) profilogram, (c) and (d) SEM-images.

4. RRAM element prototype fabrication

The lower contact topology of the resistive memory element prototype was formed using focused ion beam of the scanning electron microscope (SEM) Nova NanoLab 600 (FEI, Netherlands). Then titanium film was conducted by local anodic oxidation using probe nanolaboratory; as a result, 2 nm thick titanium oxide nanoscale structures were formed. In order to produce a cross-bar array of the memristor, the top carbon contact lines were deposited on the structures by nanolithography (Fig. 3).

5. Conclusion

We have revealed that titanium oxide nanoscale structures obtained by the local anodic oxidation demonstrate the memristive behavior without electroforming process. Increasing the height of the structures leads to current decrease and increase of their high and low resistance states values.

The obtained results can be used for development of the technological processes for fabrication of the components for microelectronics, nanoelectronics, and micro- and nanosystem engineering.

Acknowledgments
The study was supported by the RFBR (projects No. 16-32-00069 mol_a, 16-29-14023 ofi_m) and Southern Federal University (internal grants No. 07/2017-02, 07/2017-26). The results were obtained using the equipment of Research and Education Center and the Center for Collective Use "Nanotechnologies" of Southern Federal University.

References
[1] Chang T-C, Chang K-C, Tsai T-M, Chu T-J and Sze S M 2016 Resistance random access memory *Materials Today* **19** 254-64
[2] Ye C, Wu J, He G, Zhang J, Deng T, He P, Wang H 2016 Physical Mechanism and Performance Factors of Metal Oxide Based Resistive Switching Memory: A Review *Journal of Materials Science & Technology* **32** 1-11
[3] Pan F, Gao S, Chen C, Song C and Zeng F 2014 Recent progress in resistive random access memories: Materials, switching mechanisms, and performance *Materials Science and Engineering* R **83** 1-59
[4] Sieu D and Ramanathan S 2011 Adaptive oxide electronics: A review *J. Appl. Phys.* **110** 071101
[5] Avilov V I, Ageev O A, Kolomiitsev A S, Konoplev B G, Smirnov V A and Tsukanova O G 2014 Formation of a memristor matrix based on titanium oxide and investigation by probe-nanotechnology methods *Semiconductors* **48** 1757-62
[6] Ageev O A, Smirnov V A, Solodovnik M S, Rukomoikin A V and Avilov V I 2012 A Study of the formation modes of nanosized oxide structures of gallium arsenide by local anodic oxidation *Semiconductors* **46** 1616-21
[7] Chen A 2016 A review of emerging non-volatile memory (NVM) technologies and Applications *Solid-State Electronics* **125** 25-38
[8] Ageev O A, Avilov V I, Smirnov V A, Solodovnik M S and Tsukanova O G 2015 Studying the modes of nanodimensional surface profiling of gallium arsenide epitaxial structures by local anodic oxidation *Nanotechnologies in Russia* **10** 214-9
[9] Linggang Z, Jian Z, Zhonglu G and Zhimei S 2015 An overview of materials issues in resistive random access memory *J Materiomics* **1** 285-95
[10] Avilov V I, Ageev O A, Blinov Y F, Konoplev B G, Polyakov V V, Smirnov V A and Tsukanova O G 2015 Simulation of the formation of nanosize oxide structures by local anode oxidation of the metal surface *Technical Physics* **60** 717-23
[11] Siles P F, Archanjo B S, Baptista D L, Pimentel V L and Joshua J 2011 Nanoscale lateral switchable rectifiers fabricated by local anodic oxidation *J. Appl. Phys.*, **110** 024511
[12] Klimin V S, Solodovnik M S, Smirnov V A, Eskov A V, Tominov R V and Ageev O A 2016 *Proceedings of SPIE* **10224** 102241Z
[13] Acharyya D, Hazra A and Bhattacharyya P 2014 A journey towards reliability improvement of TiO_2 based resistive random access memory: A review *Microelectronics Reliability* **54** 541-60

IOP Conf. Series: Materials Science and Engineering **256** (2017) 012002 doi:10.1088/1757-899X/256/1/012002

Optomechanical coupling between AFM cantilever and semiconductor laser

P A Alekseev[1,*], V A Sharov[1], A O Mikhaylov[1], M S Dunaevskii[1] and A Baranov[2]

[1] Ioffe Institute, Saint-Petersburg, 194021, Russia
[2] Institut d'Electronique du Sud, UM 2/CNRS, Montpellier, 34095, France

*npoxep@gmail.com

Abstract. Optomechanical coupling between cantilever of the atomic force microscope (AFM) and a semiconductor laser was studied. When an AFM cantilever is oriented parallel to the surface of the emitting semiconductor laser, the system forms an optomechanical resonator. The optomechanical coupling due to photothermal effect was shown in room conditions. By optomechanical cooling the effective temperature of Si_3N_4 cantilever was reduced down to 80 K. Amplitude-to-phase conversion in the studied cavity was demonstrated.

1. Introduction

In a system consisting of optical cavity and mechanical resonator an optomechanical coupling can be realized [1]. Usually, optomechanical resonator is formed by optical cavity with movable mirror clamped by the spring. Light stored in the optical cavity due to a radiation pressure moves the mirror. Then the mirror is shifted and the cavity will be detuned with simultaneous reducing of the radiation pressure, which returns the mirror to the initial position.

Detuning of the laser from the cavity resonance allows one to control the damping rate of the mirror oscillations. For the red-detuned case (laser frequency lower than the cavity resonance) mechanical oscillations of the mirror will be damped, in other words the mirror will be "cooled" [2]. For the blue-detuned case the damping will be negative and the mirror will be "heated" with arising of parametric self-oscillations. Optomechanical coupling opens the way to cool the macroscopic mirror down to its ground state, where the quantum mechanics describes the motion of the mirror [3]. Besides the cooling, the optomechanical resonators increase the sensitivity in displacement and mass measurements and find a place in a plenty of other novel devices [4].

In atomic force microscopy (AFM), optomechanical coupling can be implemented in quality factor "Q" control of the cantilever oscillations. The cooling of the cantilever down to 18 K was achieved [5]. The cantilever served as a movable mirror and a semi-reflecting cleavage of the optical fiber – as the counterpart mirror. The optomechanical coupling was realized due to photothermal (bolometric) effect. Recently, we have studied Fabri-Perot resonances between AFM cantilever and semiconductor laser cleavage [6]. The cantilever was inclined by 20 degrees with respect to the cleavage plane and the optomechanical coupling was not observed. The strongest photothermal effect was observed on Si_3N_4 cantilever due to its low thermal conductivity.

The aim of this work was to study an optomechanical coupling between AFM cantilever and semiconductor laser in room conditions. In addition, the influence of the cantilever material was studied.

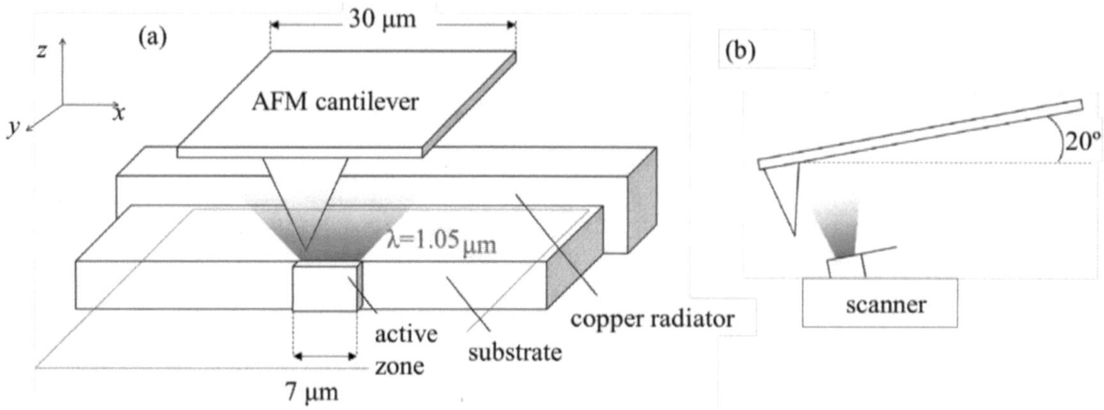

Figure 1. (a) Scheme of experimental setup for the measurements. (b) Scheme of optical cavity formed by cantilever and laser surface.

2. Experimental section

Measurements were performed using Ntegra-AURA (NT-MDT) scanning probe microscope in room conditions. We tested AFM probes fabricated of different materials: (1) silicon "Nanosensors" PPP-FMAu probes without tip coating and (2) silicon nitride "Bruker" ORC-8 probes with 30 nm thick Au tip coating. Both rectangular shaped cantilevers had the first resonance frequency f_1 ~70 kHz.

Figure 1 shows a scheme of experimental setup. The cantilever was oriented parallel to the cleavage surface of a semiconductor laser. For this purpose, the laser was mounted on a special holder with inclined surface. The inclination angle (20 degrees) was equal to that of cantilever with respect of xy scanning plane. This setup forms optical cavity with parallel mirrors: the laser surface and the cantilever surface on the tip side.

We have used semiconductor stripe laser with Fabry-Perot resonator radiating at λ=1.05 μm. Width of the laser stripe was 7 μm and the laser operated in the single mode regime. Optical power was varied in 10-100 mW range depending on the electrical pumping current (threshold current I_{th}=25 mA). During the measurements, AFM cantilever was moved with respect to laser surface by xyz scanner and a phase shift of the cantilever oscillations was recorded.

3. Results and discussion

3.1. Optomechanical coupling due to photothermal effect

In the optomechanical cavities, the optomechanical coupling can be realized by either light radiation pressure or photothermal effect. These effects have different characteristic times [7]. For the photothermal effect, the whole cantilever should be heated by laser radiation. Heated cantilever changes its stiffness due to temperature-sensitive Young's modulus [6]. Since the thermal diffusivity coefficient of silicon is of ~1 cm^2/s, the characteristic time for the silicon cantilever heating will be of 1 msec. For the light radiation pressure such a time is many orders shorter ~1 ns. Therefore, the mechanism of optomechanical coupling can be distinguished by time resolved experiment.

Figure 2 shows an absolute phase shift of the silicon cantilever oscillations during the increasing of a z-distance between cantilever and laser surface (mirrors in the optical cavity). For cantilever velocity of 7 μm/s, the phase shift dependence exhibited a sinusoidal curve specific for Fabry-Perot resonances with period corresponding to $\lambda/2$ (Fig. 2a). Such behavior has been observed earlier even when the cantilever was not parallel to the laser surface [6]. However, velocity decreasing down to 1.4 μm/s transforms the phase curve shape from "sinusoidal" to asymmetrical one, but still with periodical z-dependence (Fig. 2b). The similar curve shape was observed earlier [5] as a clear signature of the optomechanical coupling. Since the optomechanical coupling arises only for relatively slow change of the distance between mirrors, one can declare a photothermal effect responsible for the coupling.

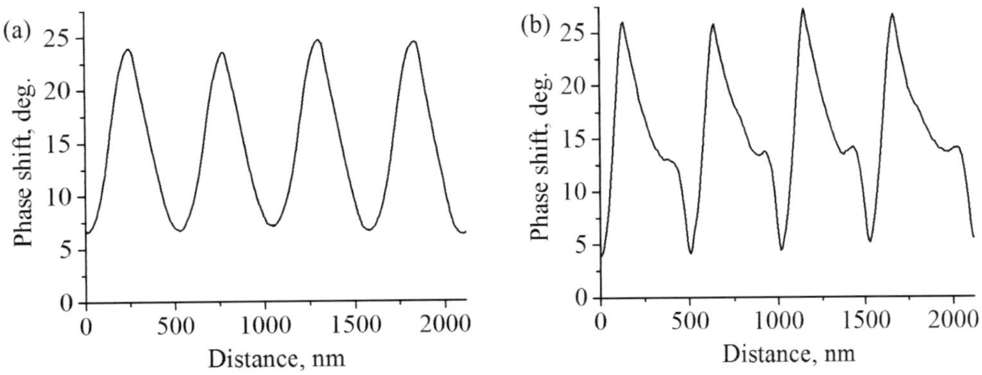

Figure 2. Dependence of the phase shift of cantilever oscillations on cantilever-laser surface z-distance for different velocities of cantilever motion along z-axis: (a) 7 μm/s, (b) 1.4 μm/s.

3.2. Optomechanical cooling of silicon and silicon nitride cantilevers

The optomechanical cooling of the cantilever was studied. For this purpose, the optical cavity was red-detuned by *z*-moving of cantilever to a proper position. Experiment was performed as follows: thermal noise spectrum of the cantilever was measured with and without laser pumping. From the obtained spectra, the resonance frequencies ω, $\omega_{\rm eff}$, and damping factors Γ, $\Gamma_{\rm eff}$ were extracted for non-lasing and lasing cases, respectively. The effective temperature of the "cooled" cantilever $T_{\rm eff}$ can be expressed as follows [5]:

$$T_{\rm eff} = T\frac{\omega^2}{\omega_{\rm eff}^2}\frac{\Gamma}{\Gamma_{\rm eff}} \tag{1}$$

where T is the room temperature.

The calculated values of $T_{\rm eff}$ thermal noise and spectra obtained with different laser pumping for Si and Si_3N_4 cantilevers are presented in Figures 3a and 3b, respectively. It is seen that the cooling of Si_3N_4 cantilever is more effective than Si one. For the Si_3N_4 cantilever, effective temperature decreases down to 80 K at 40 mA pumping current, while for the Si one even at 100 mA pumping the temperature decreases only to 170 K. Such a distinction is connected with different impacts of the photothermal effect on Si and Si_3N_4 cantilevers. Previously, we have shown a better photosensitivity of Si_3N_4 cantilever. Interestingly, in spite of the effective absorption of laser radiation by silicon, Si_3N_4 showed better results due to its lower thermal conductivity [6]. The lower thermal conductivity leads to effective optomechanical cooling; however, increasing of laser pumping current above 40 mA heated the Si_3N_4 cantilever to more than 80 K. Thus, the minimum reachable temperature of cooling at our setup was 80 K for Si_3N_4 cantilever.

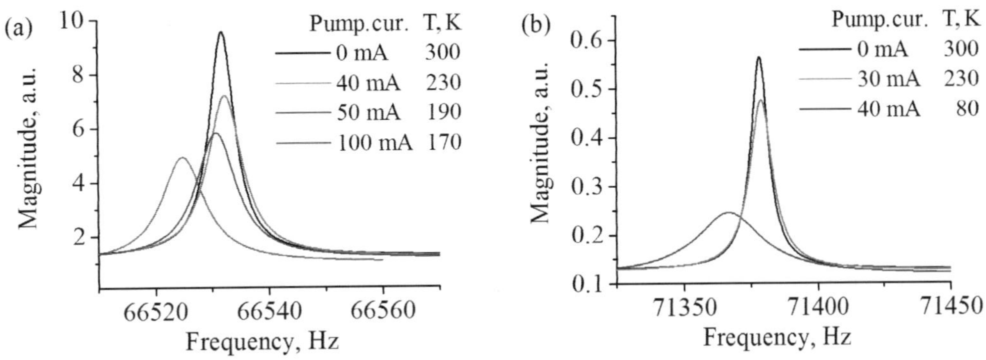

Figure 3. Thermal noise spectra of cantilever oscillations obtained with different laser pumping for (a) silicon cantilever, (b) Si_3N_4 cantilever.

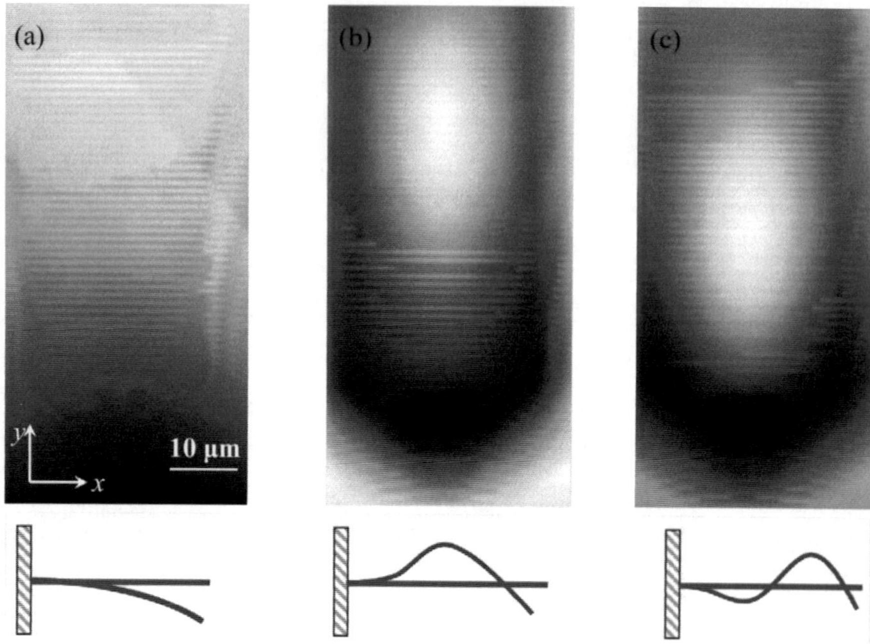

Figure 4. Phase images of the Si_3N_4 cantilever oscillations obtained while scanning semiconductor laser at various resonance frequencies: (a) f_1, (b) f_2, (c) f_3. At the bottom – flexural cantilever modes at the correspondent resonance frequencies.

3.3. XY scanning of the optomechanical cavity

In the previous section, the cantilever was moved only in *z*-direction. However, AFM provides a cantilever movement in *xy* directions (100x100 μm^2). Figure 4 shows phase images obtained while scanning a semiconductor laser in *xy* directions. The scanning was performed at different resonance frequencies of cantilever (f_1, f_2, f_3 – Figs. 4(a)-(c), respectively).

For various resonance frequencies, the bending profiles of the oscillating cantilever are different. For example, at the first resonance frequency an unclamped end of the cantilever moves up and down, and the bending profile changes monotonically. For the higher resonances the bending profiles become more complex (see bottom part of the Figure 4). It should be noted that such a bending profile can be obtained by measuring a magnitude of oscillations across the cantilever length.

Optomechanical cavity is a convenient tool of optical modulation and phase-to-amplitude conversion [4, 8]. Oscillation amplitude of a movable mirror changes the phase of optical oscillations in the cavity and vice versa. In the Figure 4, phase images of the mirror (cantilever) oscillations correlate with corresponding bending profiles. Thus, the bending profiles at different resonance frequencies can be obtained by measuring the phase image of the cantilever oscillations. Apparently, this effect can be used for measuring the bending profiles of different 1D and 2D flexural nanostructures.

4. Conclusions

To conclude, an AFM cantilever placed parallel to the surface of the mirror of the semiconductor laser forms an optomechanical cavity in room conditions. It was shown that optomechanical coupling arised due to photothermal effect. The optomechanical cooling of the AFM cantilever was studied for Si and Si_3N_4 cantilevers. The effective temperature of the Si_3N_4 cantilever was reduced down to 80 K; apparently, this value is close to the minimum reachable in our experimental setup. Scanning in the *xy* directions revealed the similarity of the phase images of cantilever oscillations at different resonance frequencies with the corresponding bending profiles of the cantilever.

Acknowledgments
P.A.A., V.A.S., and A.O.M. acknowledge support by the Presidential Grant MK-7001.2016.2. M.S.D. acknowledges support by RFBR grant (research project № 17-52-16029).

References
[1] Aspelmeyer M, Kippenberg T J and Marquardt F 2014 Cavity optomechanics *Rev. Mod. Phys.* **86** 1391–452
[2] Kippenberg T J 2010 Cavity Optomechanics *Science* **321** 1172-6
[3] Chan J, Alegre T P M, Safavi-Naeini A H, Hill J T, Krause A, Groeblacher S, Aspelmeyer M and Painter O 2011 Laser cooling of a nanomechanical oscillator into its quantum ground state *Nature* **478** 18
[4] Metcalfe M 2014 Applications of cavity optomechanics *Appl. Phys. Rev.* **1** 31105
[5] Metzger C H and Karrai K 2004 Cavity cooling of a microlever. *Nature* **432** 1002-5
[6] Dunaevskiy M, Dontsov A, Alekseev P, Monakhov A, Baranov A, Girard P, Arinero R, Teissier R and Titkov A 2015 Apertureless scanning microscope probe as a detector of semiconductor laser emission *Appl. Phys. Lett.* **106** 171105
[7] Tröger L and Reichling M 2010 Quantification of antagonistic optomechanical forces in an interferometric detection system for dynamic force microscopy *Appl. Phys. Lett.* **97** 213105
[8] Villar A S 2008 The conversion of phase to amplitude fluctuations of a light beam by an optical cavity *Am. J. Phys.* **76** 922-9

IOP Conf. Series: Materials Science and Engineering **256** (2017) 012003 doi:10.1088/1757-899X/256/1/012003

Reducing the influence of the surface roughness on the hardness measurement using instrumented indentation test

I Maslenikov[1,*], A Useinov[1], A Birykov[1] and V Reshetov[2]

[1]FSBI TISNCM, Troitsk, Moscow, Russia
[2]NRTU MEPhI, Moscow, Russia

[*]i.i.maslenikov@gmail.com

Abstract. The instrumented indentation method requires the sample surface to be flat and smooth; thus, hardness and elastic modulus values are affected by the roughness. A model that accounts for the isotropic surface roughness and can be used to correct the data in two limiting cases is proposed. Suggested approach requires the surface roughness parameters to be known.

1. Introduction

Over past several decades, instrumented indentation technique has widely been used for the materials mechanical properties measurement at the submicron and nanoscale length scale level [1]. According to this method, the hardness value is calculated as a relation of applied normal load F to the contact area A_c achieved during the corresponding indenter penetration into the sample surface. The contact area is not measured directly, but is defined by the so-called area function $A_c = f(h_c)$, where h_c is a contact depth estimated during experiment from the loading-unloading diagram. Typically, area function is calibrated on a reference sample using the series of indentations.

The method requires sample preparation, demanding that the surface has to be flat and smooth. In practice, this condition is considered to be met when the indentation depth is bigger than $20 \cdot R_a$, where Ra is the arithmetic average of the roughness profile. Such a criterion gives an uncertainty less than 5% [2]. Violating this rule leads to incorrect estimation of the contact area and, consequently, changes measured hardness value.

However, some samples, for example thin rough films and coatings, do require the indentation depth to be small, comparable to R_a and apparently smaller than $20 \cdot R_a$. Even apart from such examples, each sample roughness contributes to the measured values of hardness and elastic modulus, not only increasing the data scattering, but also biasing the mean values. Therefore, the proper model for the error estimation and correction should be useful.

2. Existing model and current approach

The idea to make an account for the roughness is to find the cases, where the interaction with rough surface can be reduced to the smooth contact but with corrected value of the contact area. A simple idea for the rough surface with relatively "frequent" peaks is suggested in [3]. The approach implies that the number of peaks being squeezed beneath the indenter tip is big enough and the mean surface slope is bigger than the indenters pyramid angle. The last assumption leads to the fact that the initial contact occurs roughly at $dz \sim 2.5 \cdot R_a$ above the mean surface line (for the normal-distributed heights). Indenter flatters the protruding asperities to the mean line and situation reduces to the flat smooth surface contact with the only correction $h_c \to h_c - dz$. The contact area A_c is recalculated according to the calibrated contact area function.

New approach proposed in the current work extends the above idea and considers the surfaces with the range of average slopes that are not necessarily bigger than the apex angle of the indenting pyramid. Moreover, another limiting case of a shallow contact with the slope of the single roughness asperity is also considered. The last case reduces to the contact with smooth inclined surface.

To create such kind of model, one needs to describe the surface spatial frequencies. This can be done either by using a PSD (power spectral density) or an autocorrelation function. In the current approach, we assumed a Gaussian autocorrelation function. The correlation tensor for the 2D-surface was defined in the following way:

$$C_{x_i,y_i,x_j,y_j} \sim \left(\exp\left(-\frac{(x_i - x_j)^2 + (y_i - y_j)^2}{\sigma_{xy}^2} \right) + 10^{-3} \left(1 - \frac{|x_i - x_j|}{L_x} \right) \left(1 - \frac{|y_i - y_j|}{L_y} \right) \right)$$

Normalizing constant was chosen to be one unit; the overall 2D array was renormalized afterwards to have some certain value of standard deviation σ_z for the height distribution. The last term in this expression was added to make the tensor condition number small enough, so that the future numerical calculation is possible.

To generate a 2D surface we've used a tensor A package available for R software [4]. Cholevsky decomposition was used to find a tensor $L_{a,b,i,j}$ so that

$$L_{i,j,a,b} L_{k,m,a,b} = C_{i,j,k,m} \qquad (1)$$

For the uncorrelated array of normal-distributed random numbers $Z0_{i,j}$, the following transformation gives normally distributed correlated array:

$$L_{i,j,a,b} \cdot Z0_{i,j} = Z_{a,b} \qquad (2)$$

7500 different $Z_{a,b}$ arrays were generated. Each of them has 128x128 points, σ_{xy} parameter was set to be equal to 4. Heights of each of 7500 arrays were rescaled to have different ratios of σ_z/σ_{xy}. 26 values of σ_z/σ_{xy} in the range of [0.02, 6] were used.

3. Contact point bias correction

For each array, an approaching of Berkovich type three-sided pyramid till the initial contact was modeled. Wolfram Mathematica software was used. The coordinate dz of the initial contact from the mean surface plane was calculated. The probability density functions (PDFs) of normalized value of dz/σ_z are presented in the Figure 1.

As it follows from the figure for surfaces with the "small slopes" i.e. $\sigma_z/\sigma_{xy} = 0.02 \ldots 0.2$, average value of dz is almost equal to zero, which means that depth measurements for such kind of surface are not biased. In this case, the error in hardness measurements more likely comes from the local slope of surface; this effect will be considered later.

As the ratio of σ_z/σ_{xy} grows, bias in the average value of $< dz >/\sigma_z$ becomes evident. This corresponds to the situation, when the initial contact occurs between the surface and pyramid's face instead of the pyramid tip. To characterize this effect, the center of mass of each PDF was calculated. Corresponding dependence of $< dz >/\sigma_z$ vs σ_z/σ_{xy} is presented in Figure 2.

Interpolation function $f(\sigma_z/\sigma_{xy})$ of $< dz >/\sigma_z$ versus σ_z/σ_{xy} dependence can be used to correct the indentation data. The correction implies recalculation of data with the following change: $h_c \to h_c - \sigma_z \cdot f(\sigma_z/\sigma_{xy})$. As it follows from the Figure 2, in the case of $\sigma_z/\sigma_{xy} \gg 1$ the value of $< dz >$ is shifted to $3\sigma_z$, i.e. corresponds to "three sigma rule" and to the case considered in work [3].

IOP Conf. Series: Materials Science and Engineering **256** (2017) 012003 doi:10.1088/1757-899X/256/1/012003

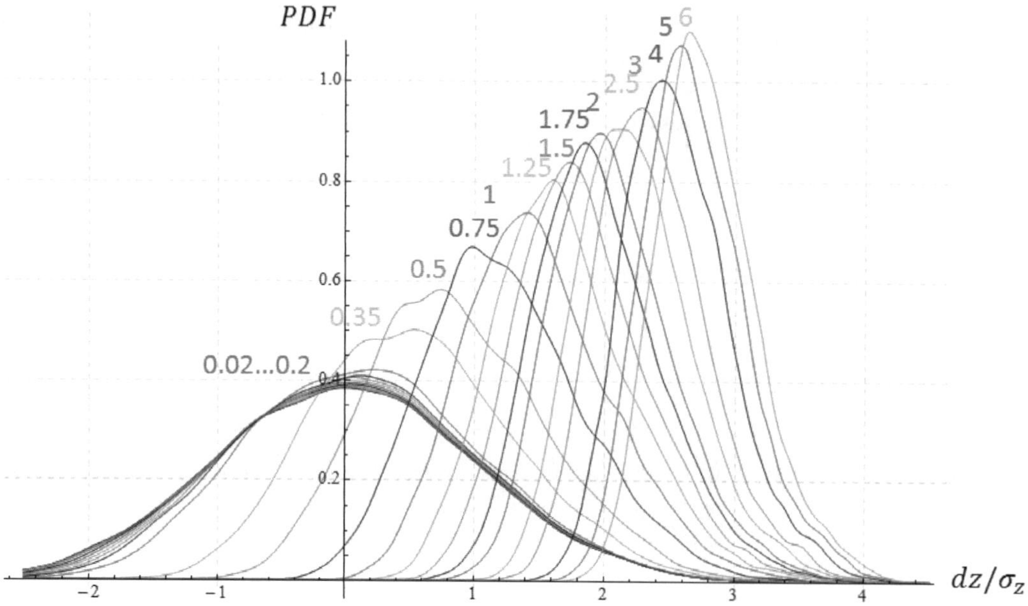

Figure 1. PDFs of normalized shift of the contact point dz/σ_z for the different values of σ_z/σ_{xy}.

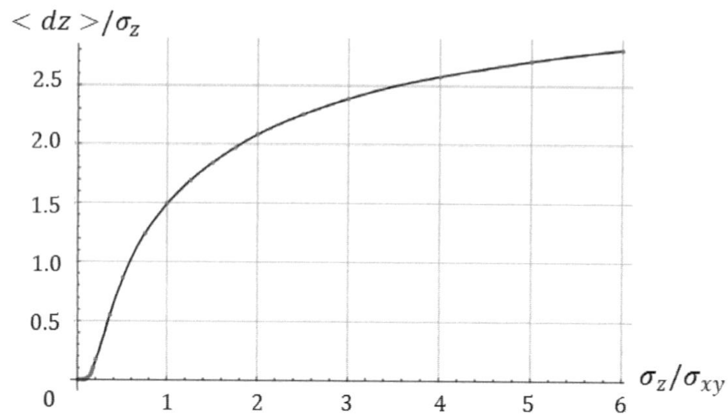

Figure 2. Average normalized contact point shift vs σ_z/σ_{xy} parameter.

4. Average slope correction

In case of small σ_z/σ_{xy} values (i.e. $\sigma_z/\sigma_{xy} \lesssim 0.1$) in average the initial contact occurs on the surface mean plane, so that the value $< dz >$ is equal to zero, but still there is an error in hardness and elastic modulus measurements that comes from the local slope.

To make an account for the described effect, let us consider the Berkovich pyramid contacting the inclined surface. To fix the idea, let us suppose that the pyramid's height is aligned along the vertical direction and the surface is inclined so that its normal has the angles φ and θ in spherical coordinates: φ is calculated from the OX axis, the pyramid edge being aligned along OY axis; θ is calculated from the pyramid height, pyramid's tip looking down.

In this case, it can be shown that for the given depth projected area of contact A_p is related to the flat surface contact area (for the same depth) A_0 according to the following relation:

$$\frac{A_p}{A_0} = \frac{cot[\alpha]^3}{(cot[\alpha]+2sin[\varphi]tan[\theta])(cot[\alpha]^2-2cot[\alpha]sin[\varphi]tan[\theta]-(1+2cos[2\varphi])tan[\theta]^2)} \quad (3)$$

where α is the angle between pyramid's height and face. As long as the surface is considered isotropic, it is necessary to average this function across all φ values:

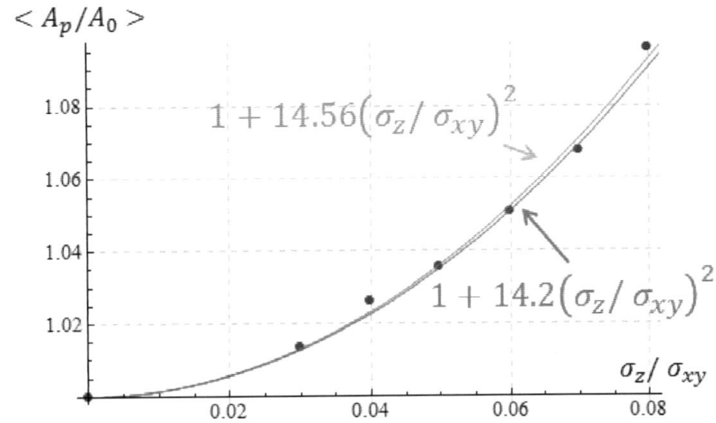

Figure 3. Correction factor for contact area in the case of single-asperity indentation: analytical expression (red line) and numerical experimental verification (green line).

$$< \frac{A_p}{A_0} > = \frac{1}{2\pi}\int_0^{2\pi} \frac{A_p}{A_0}(\varphi,\theta)d\varphi \quad (4)$$

Up to the second-order terms, the integration gives:

$$< \frac{A_p}{A_0} > \approx 1 + 3tg(\alpha)^2 \tan(\theta)^2 \quad (5)$$

For the Berkovich indenter $\alpha = 65.3^o$, which leads to the following relation:

$$< \frac{A_p}{A_0} > \approx 1 + 14.2\tan(\theta)^2 \quad (6)$$

To show the way to correct the data with the obtained equation, the following comparison with the numerical simulation is useful. We have used 300 numerically simulated surfaces for each of the 6 different values of σ_z/σ_{xy} in the range $0 < \sigma_z/\sigma_{xy} < 0.1$. Penetration of Berkovich type pyramid into each surface was modeled up to the depth of $\sigma_z/75$. The ratio of $\frac{A_p}{A_0}$ has been averaged across four depth values for each surface and then across all of the surfaces for the given σ_z/σ_{xy} value. The resulted points are presented in Figure 3 by dots. The experimental points were fitted with the parabola curve, which led to the following relation: $< A_p/A_0 >= 1 + 14.56\,(\sigma_z/\sigma_{xy})^2$.

The coincidence between two curve leads to the conclusion that the parameter σ_z/σ_{xy} is simply an equivalent of the $\tan(\theta)$ and the experimental data can be corrected by multiplying each of the calculated area value by the factor of $1 + 14.2\,(\sigma_z/\sigma_{xy})^2$.

5. The use of the model for the experimental data

The model was tested on the several polycarbonate samples with different roughness. The rough surface was obtained using the sandpaper with the different grit parameters: 320, 500, 800, and 1200. Statistical parameters of the surface were obtained using AFM Ntegra Prima with installed both upper and lower piezo-scanners allowing the scanning area up to 200x200 μm^2. The σ_z values were determined from the RMS value, σ_{xy} values were determined from the width of autocorrelation function base. Some auto-correlation functions appeared to be quite antisymmetric, so all widths were averaged across two perpendicular directions.

A series of indentations was performed on each sample, load value and number of indentation are given in Table 1. NanoScan 4D instrumented indentation tester was used.

The hardness value without correction is calculated as an average value from all the indentations: $\langle H \rangle = \frac{1}{N}\sum_i H_i$. According to the model described above, corrected value of hardness was calculated as follows: $H_{corr} = F/A_c(\frac{1}{N}\sum_i h_{c_i} - \sigma_z \cdot f(\sigma_z/\sigma_{xy}))$, where $A_c(h_c)$ is the indenter area function. For the sample with grit parameter 320 and force 50 mN, the averaged value $\langle H \rangle = 0.15$ GPa was even closer to the smooth surface value of 0.21 GPa than the corrected value, which is 0.14 GPa. Such a difference can be explained by the fact that here we compared average hardness with hardness calculated from the average contact depth. This was done intentionally, because namely the average hardness is value that is reported for the substance. Calculation of non-corrected value from the mean contact depth $F/A_c(\frac{1}{N}\sum_i h_{c_i})$ for this series would give the value of 0.13 GPa, which is farther from the "true" value of 0.21 GPa.

Table 1. Rough surface parameters and hardness measurements results.

Sample grit parameter	Amount of indents	σ_z, nm	σ_{xy}, nm	σ_z/σ_{xy}	F, mN	$\langle H \rangle$, GPa (non-corrected)	H_{corr}, GPa (corrected)
180	101	7382	27380	0.27	100	0.1	0.18
	99				1400	0.16	0.19
320	100	1770	9720	0.18	100	0.14	0.14
	100				1400	0.18	0.18
320	64	1250	8558	0.15	10	0.15	0.15
	43				50	0.15	0.14
	22				200	0.17	0.19
500	47	950	4327	0.22	10	0.16	0.2
	45				50	0.2	0.23
	12				200	0.19	0.21
800	100	1104	4380	0.25	100	0.14	0.15
	100				1400	0.16	0.17
1200	52	500	2806	0.18	10	0.2	0.21
	26				50	0.19	0.2
	16				200	0.19	0.2

6. Conclusion

Roughness biased value of hardness can be corrected using an information about the surface topography, namely the parameters that describe surface auto correlation function and height distribution. Based on these parameters, the proposed model can be used to correct the measured hardness value in case of shallow indentation (imprint size much smaller than size of autocorrelation function) and in case of deep indentation (several local peaks are covered by the contact area). In the first case, an analytical equation is suggested, which is used to make a multiplicative correction to the mean value of the measured area. The proposed equation is in good agreement with the numerical simulation result. The approach for the second case, which implies the correction for the mean contact depth, was used to correct real experimental value. A qualitative agreement between the corrected and real values of hardness was achieved. It has been shown that the proposed model allows reaching significant improvement of the accuracy of the hardness value measured using instrumented indentation test.

References

[1] Oliver W C and Pharr G M 2004 Measurement of hardness and elastic modulus by instrumented indentation: Advances in understanding and refinements to methodology *J. Mater. Res.* **19** 3-20
[2] Fischer-Cripps A C 2011 *Nanoindentation* (New York: Springer-Verlag)
[3] Kim J-Y, Lee J-J, Lee Y-H, Jang J and Kwon D 2006 Surface roughness effect in instrumented indentation: A simple contact depth model and its verification *J. Mater. Res.* **21** 2975-8
[4] https://www.r-project.org/

IOP Conf. Series: Materials Science and Engineering **256** (2017) 012004 doi:10.1088/1757-899X/256/1/012004

Multi-frequency data analysis in AFM by wavelet transform

V Pukhova[1,*] and G Ferrini[2]

[1]Department of Quantum Electronics and Optoelectronic Devices, Saint Petersburg Electrotechnical University "LETI", Saint Petersburg, 197376, Russia
[2]Interdisciplinary Laboratories for Advanced Materials Physics (I-LAMP) and Dipartimento di Matematica e Fisica, Università Cattolica del Sacro Cuore, Brescia, 25121, Italy

[*]vmpukhova@etu.ru

Abstract. Interacting cantilevers in AFM experiments generate non-stationary, multi-frequency signals consisting of numerous excited flexural and torsional modes and their harmonics. The analysis of such signals is challenging, requiring special methodological approaches and a powerful mathematical apparatus. The most common approach to the signal analysis is to apply Fourier transform analysis. However, FT gives accurate spectra for stationary signals, and for signals changing their spectral content over time, FT provides only an averaged spectrum. Hence, for non-stationary and rapidly varying signals, such as those from interacting cantilevers, a method that shows the spectral evolution in time is needed. One of the most powerful techniques, allowing detailed time-frequency representation of signals, is the wavelet transform. It is a method of analysis that allows representation of energy associated to the signal at a particular frequency and time, providing correlation between the spectral and temporal features of the signal, unlike FT. This is particularly important in AFM experiments because signals nonlinearities contains valuable information about tip-sample interactions and consequently surfaces properties. The present work is aimed to show the advantages of wavelet transform in comparison with FT using as an example the force curve analysis in dynamic force spectroscopy.

1. Introduction

Scanning probes used in Atomic Force Microscopy (AFM) [1] are micromechanical oscillators and the theory of AFM dynamics is based on the analysis of the oscillating frequencies of beam resonators. Interacting cantilevers in AFM experiments generate non-stationary, multi-frequency signals consisting of numerous excited flexural and torsional modes and their harmonics.

In conventional dynamic AFM methods the excitation and detection of the cantilever oscillations are performed at a single frequency. While the tip is interacting with the surface, higher frequency components in the cantilever dynamics arise caused by the nonlinearities in the interaction force. These components carry information about the sample properties. The existence of these additional components in the detected signal has been known for several years, but their role in the cantilever dynamic, spatial resolution, and sensitivity to material properties was not appreciated.

The oscillations of a non-interacting cantilever driven at frequency ω are nearly perfect sinusoids that can be described as $Acos(\omega t + \varphi)$, where A is the amplitude of the oscillations and φ is the phase lag with respect to the driver. When the tip of the excited cantilever comes in close vicinity of the sample surface, it starts experiencing distortions because of forces acting on the tip and this leads to the excitation of higher harmonics [2, 3]. Thus, the frequency spectrum of the higher harmonics contains detailed information on the nonlinearities of the tip-sample interaction forces [4, 5]. The contribution of the components containing higher harmonics in the cantilever motion can be described as $\sum_{n=1}^{\infty} a_n \cos(n\omega t + \varphi_n)$, where a_n is the amplitude (note that it decreases as $1/n^2$) and φ_n is the phase shift of the n-harmonic, n is an integer number. In general, the higher harmonics provide enhanced lateral resolution, because the frequency shift due to the tip-sample interaction is amplified by a factor of n for the n-th harmonic [6, 7]. The higher harmonics amplitudes are not influenced by the surface topography, but by the borders of patterns, where the geometry of the tip-sample contact varies. Because of this, the excited higher harmonics provide high sensitivity to the surface roughness [6]. In the case of heterogeneous samples, the amplitudes of the higher harmonics change with respect to the mechanical properties of the regions. They can also be used to probe the elastic properties and dissipative forces [8, 9] and to perform the compositional mapping in air [10] and liquid environments [11, 12].

The higher harmonics should be distinguished from the higher modes of the cantilever. The higher harmonics are consecutive integer multiples of the fundamental cantilever resonance. The higher eigenfrequencies of the oscillating beam, however, do not form a series of consecutive integer multiples [13]. As an example, for rectangular cantilever the frequency of the second eigenmode ($6.27f$) is close to the sixth harmonic ($6f$). The values of the frequencies of the higher harmonics are close to those of the flexural eigenmodes of the cantilever. This leads to the fact that the higher harmonics act as driving forces and are able to excite the higher modes of the cantilever that are close to the integer multiples of the fundamental mode. This nonlinear modal interaction is called internal resonance [2, 6]. So, when the tip is close to the surface and is influenced by the nonlinear surface interaction forces, higher harmonics are excited. And these higher harmonics excite the oscillations of the higher eigenmodes of the cantilever due to the internal resonance. The higher modes contribute to the motion of the cantilever as $\sum_{i=1}^{\infty} b_i \cos(\omega_i t + \psi_i)$, where ω_i is the frequency, b_i is the amplitude and ψ_i is the phase shift of the i-eigenmode, i is an integer number [14]. The variations in the oscillating parameters of the higher eigenmodes are very sensitive to changes in the tip-sample interaction forces, consequently increasing the sensitivity to the mechanical, magnetic, or electric tip-sample interactions [15-17].

A number of techniques have been implemented to gain information from the tip-sample interactions, but usually the interaction of the tip with the surface is revealed by the modification of the average value of the amplitude, frequency, and phase shift over many oscillation cycles [18, 19]. The averaging techniques provide superior sensibility, allowing to probe the details of force interactions down to the molecular level [18], but they are not allowing to capture the information conveyed by the sensing tip in a single interaction, during few oscillation cycles of the cantilever. This article concentrates on the analysis technique that is able to characterize all cantilever frequency components at once, without averaging.

2. Data analysis technique

The goal of any AFM experiment is to gain information about the properties of the sample surface encoded in the dynamics of the cantilever by exploring the characteristics of the signal via signal transformation.

The most common approach to the signal analysis is to apply the Fourier transform (FT) analysis, which decomposes the signal into constituent frequencies displayed in the spectrum as resonance peaks [20]. Attributing these peaks to the frequencies that caused them in the case of stationary signals is an easy task, but it can be challenging for non-stationary signals since their spectral characteristics change over time. FT analysis gives meaningful spectra in case of stationary signals. For non-stationary and rapidly varying signals, which spectral content changes over time, such as those from interacting cantilevers in AFM experiments, FT provides an averaged spectrum, which is integrated over the whole acquisition time. Hence, to analyze this type of signals, a method that allows representing the spectral evolution in time is needed.

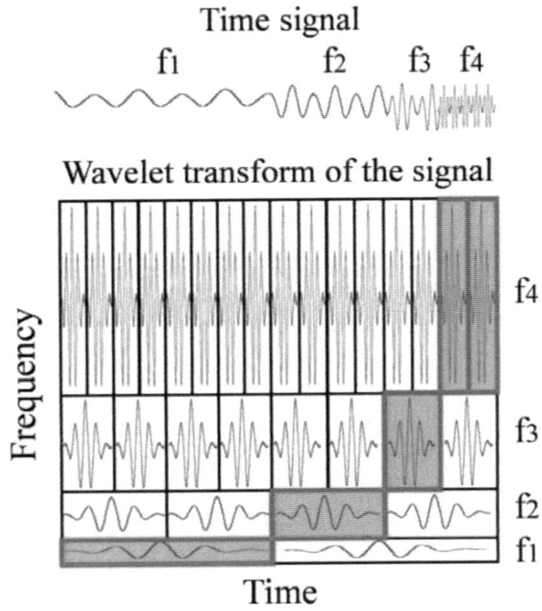

Figure 1. Principle of the WT technique.

One of the techniques allowing detailed time-frequency representation of the signals is the wavelet transform (WT) [20, 21]. WT is a method of analysis that allows representation of energy associated to the signal at a particular frequency and time, providing correlation between the spectral and temporal features of the signal. This is particularly important in AFM, because signals nonlinearities contain valuable information about tip-sample interactions and, consequently, surfaces properties [22-25].

The wavelet is an oscillating waveform function with limited support in time and zero average. The WT of the signal is the projection of the time signal onto a set of scaled and delayed wavelets; hence, the WT gives a two-dimensional time-frequency representation of a temporal signal. The original wavelet is called a mother wavelet and scaled and delayed wavelets are called daughter wavelets. The mother wavelet should be chosen according to the characteristics of the signal and the data analysis task [20].

The daughter wavelets are determined as [20]:

$$\Psi_{s,d}(t) = \frac{1}{\sqrt{s}}\Psi\left(\frac{t-d}{s}\right), \tag{1}$$

where $\Psi(t)$ is the mother wavelet, s is the positive adimensional scale parameter (stretches or compresses the mother wavelet and is connected to the frequency) and d is the delay parameter (shifts the wavelet along the time axis and is connected to the time).

The WT of the signal is determined as [20]:

$$W^f(s,d) = \int_{-\infty}^{+\infty} f(t)\Psi_{s,d}^*(t)dt = \int_{-\infty}^{+\infty} f(t)\frac{1}{\sqrt{s}}\Psi^*\left(\frac{t-d}{s}\right)dt, \tag{2}$$

where $f(t)$ is a temporal signal and $W^f(s,d)$ are wavelet coefficients that display a comparative similarity between the signal and the particular daughter wavelet function.

To illustrate the principle of the WT technique, consider the analysis of the time signal presented in Figure 1. The time signal is a sequence of several signals of different frequencies, which appear consecutively one after another. The first half of the signal is of frequency f_1, followed by the signal of frequency f_2 (quarter of the total signal duration), the next part of the signal is of frequency f_3 (eighth part of the total signal duration), and the remaining part of the signal is of frequency f_4, where $f_1 < f_2 < f_3 < f_4$. Wavelet analysis transforms the signal into a two-dimensional time-frequency representation. This two-dimensional space is orderly filled with the daughter wavelets, so that the vertical axis shows the change of the frequency, which is connected to the scale parameter, and the horizontal axis shows the change of the time, which is connected to the delay parameter, as shown in Figure 1. The wavelet transform, in this case, is carried out in the following way: each daughter wavelet from the two-dimensional time-frequency plane is compared with the corresponding part of the analyzed signal. The magnitude of the similarity between the analyzed part of the signal and the particular daughter wavelet gives the value of the wavelet coefficient, which is encoded in the time-frequency domain according to a given rule. The regions corresponding to the highest similarity between the parts of the signal and the daughter wavelets from the time-frequency plane are highlighted (Fig. 1 – red areas). According to this principle, the result of the wavelet analysis is the wavelet spectrum, schematically shown in Figure 1. From the WT spectrum it is possible to extract not only the set of frequencies constituting the signal, but also the exact frequency forming the signal at each moment.

The full spectral response of the signal can be provided, if the phase analysis is performed simultaneously with the amplitude analysis. The phase analysis is based on the cross-wavelet transform (XWT) [23].

The XWT spectrum of two signals $h(t)$ and $g(t)$ is defined as [20]:

$$W^{hg}(s,d) = W^{h}(s,d)W^{g*}(s,d), \tag{3}$$

where $W^{hg}(s,d)$ are the cross-wavelet amplitudes, $W^{h}(s,d)$ and $W^{g*}(s,d)$ are the WT of the time signals and $*$ symbolizes the complex conjugate.

The comparative phase difference between the two time signals at particular time-frequency moment (s,d) is defined as [20]:

$$\Phi^{hg}(s,d) = \Phi^{h}(s,d) - \Phi^{g}(s,d) = tan^{-1}\left(\frac{\Im(<s^{-1}W^{hg}(s,d)>)}{\Re(<s^{-1}W^{hg}(s,d)>)}\right), \tag{4}$$

where $\Phi^{h}(s,d)$ is the phase of signal $h(t)$, $\Phi^{g}(s,d)$ is the phase of signal $g(t)$, $<>$ represents the smoothing operator, $\Re$ and $\Im$ are the real and imaginary parts, respectively.

Taking into account that the signal consists of a sum of Fourier components oscillating at different frequencies, the phases of each component of the time signal can be measured, if a reference signal at the same frequency is available. It is convenient to have a *sinc* ("sinus cardinalis") function as a reference signal [23]. The *sinc* function is composed of a superposition of a number of cosines that have a maximum at zero time and oscillation frequencies in the interval between zero and maximal cut-off frequency. Since the time width of the *sinc* function is inversely proportional to its bandwidth, it gives the phase reference only near zero time. To have a phase reference at other times, the *sinc* function has to be translated along the time scale to the points of interest. Measuring by the XWT the instantaneous phase difference between the signal and the time-translated reference, *sinc* function allows forming a "phase carpet" in the time-frequency plane [23].

To illustrate the principle of the XWT technique, consider the analysis of the signal shown in Figure 2. The signal (gray line) consists of a sum of two cosines oscillating at different frequencies, $f_1 = 12$ kHz (blue line) and $f_2 = 6.5f_1 = 75$ kHz (green line), with an amplitude ratio of $A_1{:}A_2 = 3{:}1$, where A_1 and A_2 are the amplitudes of cosines components of the signal, and zero phase at zero time for both components (Fig. 2a). In Figure 2b the wavelet analysis of the signal in Figure 2a is shown. In Figure 2b the wavelet coefficients are coded in a color-scale and it can be seen that two oscillating components that combine the signal are highlighted, so the red line corresponds to the component with frequency f_1 and the green line to the one with frequency f_2. The relative phase obtained as a result of the wavelet cross-correlations between the analyzed signal and the reference *sinc* function is coded in the slope of the arrows superposed on the wavelet spectra. Black arrows correspond to the component with frequency f_1, red arrows – to the one with frequency f_2, and gray arrows show edge effect; the other frequencies are absent in the signal. Arrows pointing right indicate 0°, left – 180°, up – -90° and down – 90° of relative phase shift between the time signal and reference *sinc* function. The intermediate phases are in between these values.

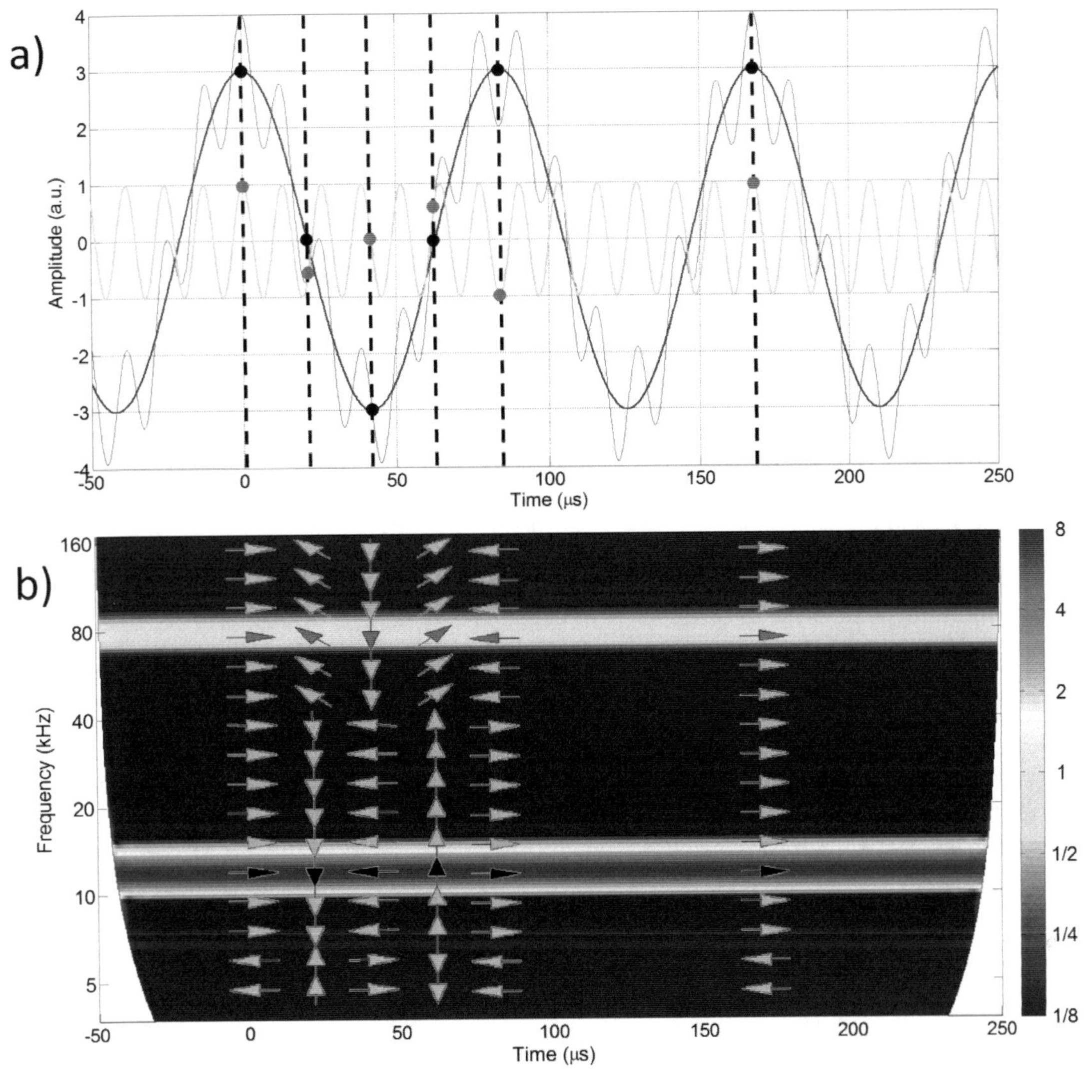

Figure 2. Principle of the XWT technique. (a) The signal (gray line) consists of a sum of two cosines oscillating at frequencies $f_1 = 12$ kHz (blue line) and $f_2 = 6.5f_1 = 75$kHz (green line), with an amplitude ratio of $A_1 : A_2 = 3 : 1$, and zero phases at time zero for both components. The black dashed lines correspond to time moments (from left to right): $0, \frac{T}{4}, \frac{T}{2}, \frac{3T}{4}, T$, and $2T$, where $T = 1/f_1$. (b) XWT analysis of the signal at time moments $0, \frac{T}{4}, \frac{T}{2}, \frac{3T}{4}, T$, and $2T$. The wavelet coefficients are coded in a color-scale. Both color-scale and frequency axes have base-2 logarithm scales. Edge artifacts of the WT are delimited by a white shade.

The black dashed lines in Figure 2a correspond to time moments (from left to right): $0, \frac{T}{4}, \frac{T}{2}, \frac{3T}{4}, T$, and $2T$, where $T = 1/f_1$, and the XWT analysis at these time moments is shown in Figure 2(b). At the zero time both oscillating components of the signal and *sinc* function are in phase; hence, the phase shifts are equal 0° (black and red arrows are pointing right). Moving the reference *sinc* function along time scale and measuring the relative phase shift between the oscillating components of the signal and *sinc* at time moments $\frac{T}{4}, \frac{T}{2}, \frac{3T}{4}, T$ we can verify that for the component with frequency f_1 the phase shifts are successively equal to 90°, 180°, -90°, and 0°, and for the component with frequency f_2 the phase shifts are successively equal to -135°, 90°, -45°, and 180°. To have both oscillating components of the signal and *sinc* function in phase, as it is at zero time, we need to move to the shortest common multiple period, i.e. $2T$ (both red and black arrows are pointing right).

3. Multi-frequency representation of the signals by wavelet transform

The AFM allows not only to investigate the topography of surfaces, but also to study the properties of surfaces by measuring tip-sample forces, commonly known as force spectroscopy. The curves obtained during force spectroscopy experiments are called force curves [26, 27].

To exemplify the data analysis technique based on the WT, a force curve is investigated, in order to show the evolution of the signal spectrum during a dynamic force spectroscopy experiment. The force spectroscopy experiment requires sequential measurements of the magnitude of the cantilever deflection due to the action of forces on its tip, while it approaches the sample surface. These measurements are generally performed at a number of fixed distances from the sample surface. When a dynamic force spectroscopy experiment is conducted, the approach of the probe to the sample surface is carried out at a constant speed, i.e. the instantaneous changes of the cantilever deflection are continuously recorded. An example of the force curve obtained by dynamic force spectroscopy with a thermally driven cantilever in ambient conditions is shown in Figure 3.

Dynamic force spectroscopy experiment has been performed on HOPG in air (relative humidity is 55%, room temperature is 296 K, atmospheric pressure) with constant approach velocity of 0.82 nm/ms and rectangular Au-coated silicon cantilever with the cantilever first free flexural eigenmode at a frequency of 11.8 kHz and spring constant equals 0.14 N/m (measured using thermal tune method applied to the first flexural mode [13]). The inverse optical lever sensitivity [28] is equal to 186.32 nm/V.

In Figure 3, the free oscillations around the zero deflection line due to the thermal excitation of the tip approaching the surface correspond to the negative part of the time scale. A sharp downward deflection (of the order of 4 nm) near zero time corresponds to the jump-to-contact (JTC) transition. The contact line, which shows the cantilever behavior, when the tip is in contact with the sample surface after the JTC transition, corresponds to the positive part of the time scale.

Analysis of the force curve (Fig. 3) by the Fourier transform is presented in Figure 4. On the Fourier power spectrum (Fig. 4a) few resonance peaks (labeled 1, 2, 3, 4) can be seen. The highest peak at a frequency of 11.6 kHz clearly corresponds to the first flexural mode. The peak number 2 has frequency of 16.1 kHz and cannot be the second flexural mode of the cantilever. The frequency of the second flexural mode should be equal to 73.9 kHz, since it is 6.27 times higher than the first one [13];

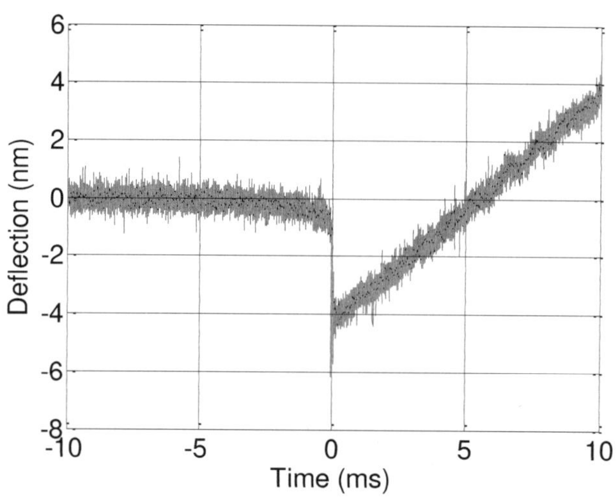

Figure 3. Deflections of the cantilever as a function of time during the dynamic force spectroscopy experiment measured by a standard optical beam deflection system.

IOP Conf. Series: Materials Science and Engineering **256** (2017) 012004 doi:10.1088/1757-899X/256/1/012004

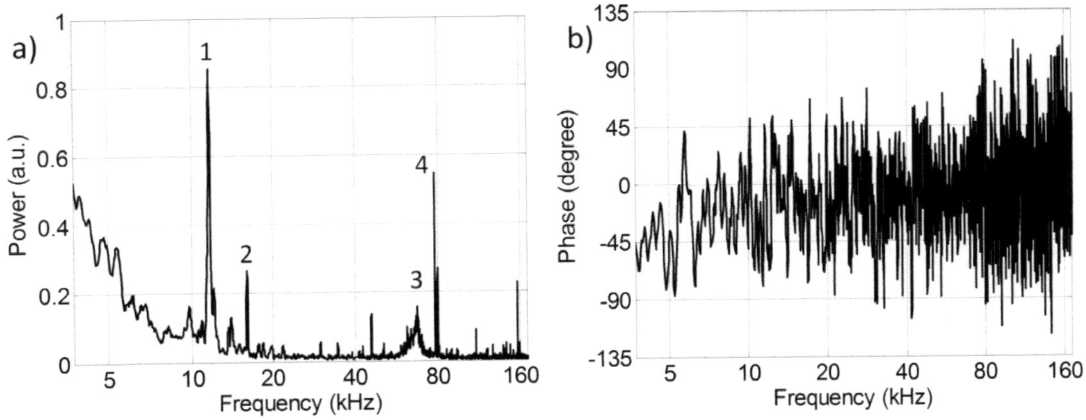

Figure 4. Fourier power (a) and phase (b) spectra of the signal shown in Figure 3. The frequency axes on both frames have base-2 logarithm scales.

hence, either peak number 3 at a frequency of 65.9 kHz, or number 4 at a frequency of 77.4 kHz, could be assigned to the second flexural mode. It is not possible to make an appropriate assignment just from the FT spectra. Moreover, it is worth noting that during the dynamic force spectroscopy experiment the cantilever eigenmodes structure is constantly changing due to the variation of external forces and boundary conditions, but Fourier spectrum in Figure 4a is the result of averaging over 20 ms, and it does not show the development of the spectrum in time. To attribute the FT spectral components of the signal, we look at the WT spectrum.

The analysis of the force curve (Fig. 3) by using the wavelet transform with the mother wavelet Morlet [20, 29] is presented in Figure 5. On the negative part of the time scale, the free oscillations of the first (bright yellow line) and the second (light blue line) flexural modes of the cantilever approaching the surface (Fig. 3) can be seen. The first flexural mode of the free cantilever on WT spectrum has a frequency of 11.9 kHz, that is down shifted before the JTC transition (here the tip approaches the surface and the tip-sample force gradient is responsible for this phenomenon). The second flexural mode has a frequency of 78.1 kHz. On the wavelet spectrum, it is easy to attribute the observed frequencies, in contrast to the Fourier spectrum.

The zero on the time scale corresponds to the JTC transition and is accompanied by several changes in the spectrum: spectral broadening around the first flexural mode and excitation in a short time interval of the second flexural mode at a frequency of 62.2 kHz. The detailed analysis of the signal around JTC transition is given in [23, 24].

On the positive part of the time scale, i.e. the oscillations of the cantilever after the JTC transition (Fig. 3), we note a resonance at a frequency of 16.3 kHz (light blue line at the bottom), slightly higher than the first free flexural mode, and a higher resonance at almost the same frequency as the second free flexural mode (light blue line at the top).These changes in spectrum are interesting because they cannot be attributed to the flexural modes of the pinned cantilever and, hence, suggest the existence of a complex interplay between the tip and the sample surface. The detailed analysis of this phenomenon on various surfaces can be found in [24].

Considering the WT spectra evolution (Fig. 5), the FT spectral components of the signal in Figure 4a can be easily attributed, since the temporal evolution of the peaks is shown as follows: peak number 2 corresponds to the frequency that appears after JTC and is a modified first free flexural mode, peak number 3 is caused by the excitation of the second flexural mode during the JTC, and peak number 3 is the second flexural mode of the cantilever.

IOP Conf. Series: Materials Science and Engineering **256** (2017) 012004 doi:10.1088/1757-899X/256/1/012004

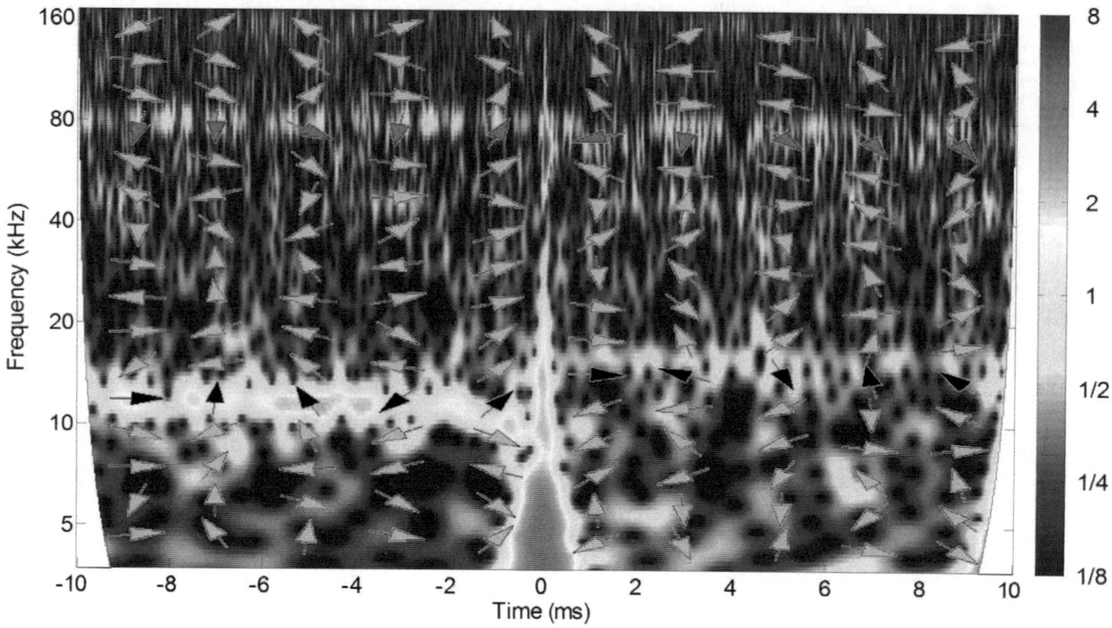

Figure 5. Wavelet transform of the signal shown in Figure 3. Wavelet coefficients are coded in color-scale. Both color-scale and frequency axes have base-2 logarithm scales. Edge artifacts of the WT are delimited by a white shade. The relative phase obtained as a result of the wavelet cross-correlations between the analyzed signal and the *sinc* function is coded in the slope of the arrows as in Figure 2. Black arrows correspond to the first flexural mode, red arrows belong to the second flexural mode, and gray arrows refer to all other frequencies.

The XWT phase measurements are coded in the slope of the arrows (Fig. 5). Black arrows correspond to the first flexural mode, red arrows belong to the second flexural mode, and gray arrows refer to all other frequencies. The measurements were carried out using the *sinc* function as a reference. In Figure 5, the XWT measurements before the JTC have been done with the time interval equals $25T$ (about 2 ms), where T is the oscillation period of the first flexural mode. In this case, the phase shift of the first flexural mode with respect to the translated *sinc* function should remain constant, if the cantilever oscillation is described by a pure sinusoid, as explained in section 2. However, it can be seen that the phase shift changes randomly over time. This is due to the fact that the observed signal is the response of the cantilever to the thermal excitation or thermal noise. The thermal excitation is a wideband random excitation constituted by molecular impacts on the cantilever by the molecules of the surrounding air in thermal agitation. As a consequence, the driving force acting on the cantilever is totally uncorrelated in time. From the XWT analysis, it appears that black and red arrows rotate randomly in time at the cantilever flexural modes, being a distinctive sign of the phase uncorrelation of the thermal excitation.

After JTC transition, cantilever is spring-coupled to the surface [24] and the resonance frequency of the first flexural mode as well as its oscillation period change (Fig. 5). XWT measurements after the JTC have been performed in the same way as for free oscillations. XWT measurements have been done with a time interval equals $30T'$ (~ 2 ms), where T' is the oscillation period of the first flexural mode after JTC. It can be seen that the relative phase shift of the first and second modes after JTC changes randomly within the analysis time frame (black and red arrows rotate). Again, XWT analysis provides an observation of the phase uncorrelation of the thermal excitation. Depth analysis is beyond the scope of the present work; however, this example shows possibility of XWT technique to access the phase of all spectral components of the signal simultaneously and at the time moments of interest, unlike FT, which gives phase spectra averaged over the entire acquisition time interval (Fig. 4b).

4. Conclusions

Wavelet analysis has been already successfully used to perform multi-frequency data analysis in AFM [22-25]; however, the whole range of its possibilities as a data analysis technique is not fully explored. With an appropriate theory, the WT can make a significant contribution to the definition of the samples surface properties that cannot be readily measured using other methods. This is ensured by the fact that the WT allows time-frequency analysis for each oscillating frequency of the cantilever, being able to extract such information without averaging. Using WT along with XWT analysis allows detecting the phases of the oscillating frequencies over time, thus providing additional valuable information on the signal.

The joint use of FT and WT provides the most comprehensive analysis of the signal [23-25], allowing to attribute the spectral components correctly, both in the wavelet spectrum and in the Fourier spectrum.

The approach discussed in this work is of great potential for the development of AFM, but it can be applied in various fields of science, where rapidly varying and non-stationary signals need to be investigated.

References

[1] Binnig G, Quate C C and Gerber C 1986 *Phys. Rev. Lett.* **56** 930
[2] Raman A, Melcher J and Tung R 2008 *Nano Today* **3** 20
[3] Garcia R and Herruzo E T 2012 *Nature Nanotech.* **7** 217
[4] Stark R W 2010 *Materials today* **13** 24
[5] Sahin O and Alatar A 2001 *Appl. Phys. Lett.* **79** 4455
[6] Balantekin M and Alatar A 2005 *Appl. Phys. Lett.* **87** 243513
[7] Hembacher S, Giessibl F J and Mannhart J 2004 *Science* **305** 380
[8] Sebastian A, Salapaka M V, Chen D J and Cleveland J P 1999 *Proc. American Control Conf. (San Diego)* vol 1 (San Diego: IEEE) pp 232-6
[9] Hu S, Howell S, Raman A, Reifenberger R and Franchek M 2004 *J. Vib. Acoust.* **126** 343
[10] Stark R W and Heckl W M 2003 *Rev. Sci. Instrum.* **74** 5111
[11] Schiener J, Witt S, Stark M and Guckenberger R 2004 *Rev. Sci. Instrum.* **75** 2564
[12] Raman A, Trigueros S, Cartagena A, Stevenson A P Z, Susilo M, Nauman E and Contera S A 2011 *Nature Nanotech.* **6** 809
[13] Butt H J and Jaschke M 1995 *Nanotechnology* **6** 1
[14] Lozano J R and Garcia R 2008 *Phys. Rev. Lett.* **100** 076102
[15] Martinez N F, Patil S, Lozano J R and Garcia R 2006 *Appl. Phys. Lett.* **89** 153115
[16] Stark R W and Heckl W M 2000 *Surf. Sci.* **457** 219
[17] Xu X and Raman A 2007 *J. Appl. Phys.* **102** 034303
[18] Giessibl F J 2003 *Rev. Mod. Phys.* **75** 949
[19] Garcia R and Perez R 2002 *Surf. Sci. Reports* **47** 197
[20] Mallat S 1999 *A Wavelet Tour of Signal Processing* (San Diego: Academic Press)
[21] Pukhova V, Gorelova E, Ferrini G and Burnasheva S 2017 *Proc. 2017 IEEE Russia Section Young Researchers in Electrical and Electronic Engineering Conf. (2017 ElConRus) (St.Petersburg)* (St.Petersburg: IEEE) 16839467
[22] Malegori G and Ferrini G 2010 *Beilstein J. Nanotechnol.* **1** 172
[23] Pukhova V, Banfi F and Ferrini G 2013 *Nanotechnology* **24** 505716
[24] Pukhova V, Banfi F and Ferrini G 2015 *Nanotechnology* **26** 175701
[25] Pukhova V, Banfi F and Ferrini G 2014 *Beilstein J. Nanotechnol.* **5** 494
[26] Butt H J, Cappella B and Kappl M 2005 *Surf. Sci. Reports* **59** 1
[27] Cappella B and Dietler G 1999 *Surf. Sci. Reports* **34** 1
[28] Higgins M J, Proksch R, Sader J E, Polcik M, Mc Endoo S, Cleveland J P and Jarvis S P 2006 *Rev. Sci. Instrum.* **77** 013701
[29] Deng Y, Wang C, Chai L and Zhang Z 2005 *Appl. Phys.* B **81** 1107

IOP Conf. Series: Materials Science and Engineering **256** (2017) 012005 doi:10.1088/1757-899X/256/1/012005

Investigation of photoconductivity of individual InAs/GaAs(001) quantum dots by Scanning Near-field Optical Microscopy

D O Filatov[1,*], I A Kazantseva[2], N V Baidus'[3], A P Gorshkov[1,3] and V P Mishkin[4]

[1] Research and Education Center for Physics of Solid State Nanostructures, Lobachevskii State University of Nizhnii Novgorod, Nizhnii Novgorod, 603950, Russia

[2] Department of Physics, Lobachevskii State University of Nizhnii Novgorod, Nizhnii Novgorod, 603950, Russia

[3] Research Institute for Physics and Technology, Lobachevskii State University of Nizhnii Novgorod, Nizhnii Novgorod, 603950, Russia

[4] Institute of Physics and Chemistry, Ogarev Mordovia State University, Saransk, Mordovia Republic 430005, Russia

*dmitry_filatov@inbox.ru

Abstract. The spatial distribution of the photocurrent in the input window plane of a GaAs-based *p–i–n* photodiode with embedded self-assembled InAs quantum dots (QDs) has been studied with the photoexcitation through a Scanning Near-field Optical Microscope (SNOM) probe at the emission wavelength greater than the intrinsic absorption edge of the host material (GaAs). The inhomogeneities related to the interband absorption in the individual InAs/GaAs(001) QDs have been observed in the photocurrent SNOM images. Thus, the possibility of imaging the individual InAs/GaAs(001) QDs in the photocurrent SNOM images with the lateral spatial resolution ~ 100 nm (of the same order of magnitude as the SNOM probe aperture size) has been demonstrated.

1. Introduction

Scanning Near-field Optical Microscopy (SNOM) is a method of Scanning Probe Microscopy destined for measuring the spatial distribution of various optical parameters of the solid surfaces with the nanometer-scale spatial resolution (much less than the wavelength of the probing light, thus overcoming the diffraction limit of the spatial resolution of the ordinary optical microscopy) [1]. Among the optical parameters mentioned above, there are optical transmittance, reflectance, luminescence, Raman scattering, etc. However, the photoconductivity (PC) of the semiconductor (and other) nanostructures under the local photoexcitation via a SNOM probe [2] is studied rarely as compared to other SNOM techniques. At the same time, the PC SNOM studies of the self-assembled semiconductor nanoislands and quantum dots (QDs) have not been reported until recently.

In Ref. [3], we have applied PC SNOM to study the photocurrent of a Si-based p^{+}–n photodiode (PD) with embedded self-assembled GeSi/Si(001) nanoislands as a function of the position of the SNOM probe in the plane of the input window of the PD. The features observed in the PC SNOM images have been attributed to the interband optical absorption in the GeSi nanoislands followed by the emission of the photoexcited holes from the nanoislands and the separation of the electron-hole pairs in the electric field of the p^+–n junction. Thus, the possibility of imaging the individual GeSi nanoislands with the lateral spatial resolution ~ 100 nm (i.e. of the same order of magnitude as the aperture size of the SNOM probe used) has been demonstrated.

In the present paper, we report on the PC SNOM investigation of the GaAs-based *p–i–n* PD with embedded self-assembled InAs/GaAs(001) QDs. The goal of the present study was to explore the capabilities of PC SNOM in the imaging of individual InAs/GaAs(001) QDs, which are much smaller (about an order of magnitude) than the GeSi/Si(001) nanoislands studied in Ref. [3].

2. Experimental details

The *p–i–n* diode structure has been grown on an n^+-GaAs(001) substrate by Low Pressure Metal Organic Vapor Phase Epitaxy (LP MOVPE) using Aixtron® AIX 200RF growth setup. The gas pressure in the reactor was ~ 100 mbar. A schematic representation of the investigated diode is shown in Figure 1. The InAs QDs were grown on the surface of a ≈ 350 nm thick intentionally undoped *i*-GaAs layer (the background electron concentration was ~ 5×10^{15} cm^{-3}) by self-assembling via Stranski-Krastanov mechanism at the substrate temperature $T_g \approx 490$ °C. The InAs QDs were capped by a ≈ 10 nm thick *i*-GaAs layer. Finally, a p^+-GaAs subcontact layer with the thickness $d_c \approx 50$ nm doped by C with the acceptor concentration ≈ 3×10^{18} cm^{-3} was deposited. The p^+-GaAs layer thickness d_c was selected so that the QDs were within the near-field region from the surface, i. e. $d_c < \lambda_{ex}/n$, where $\lambda_{ex} \sim 1$ μm is the photoexcitation wavelength and $n \approx 3.5$ is the refractive index of GaAs. A calculated equilibrium band diagram of the *p–i–n* diode structure with InAs QDs at 300 K is presented in Figure 2a. The values of the band offsets at the InAs/GaAs(001) QD interface as well as the dependencies of the quantum confined electron and hole energies on the QD sizes were taken from Ref. [4]. The charging of the QDs by holes was neglected.

The morphology of the InAs QDs were examined by Atomic Force Microscopy (AFM) in ambient air using NT-MDT® Solver Pro™ instrument in the Semicontact™ mode on the satellite sample with the surface InAs QDs grown in the same conditions as the ones in the *p–i–n* diode structure. NT-MDT® NSG-11 probes with the tip curvature radius $R_p < 10$ nm were used. The AFM data were analyzed by NT-MDT® Nova™ Image Analysis software using Grain Analysis block.

The InAs QDs in the AFM images were identified by the threshold method, and the statistical analysis of the morphological parameters of the QDs has been performed. The morphology of the QDs was characterized by the following parameters: the height h (the origin being selected at the InAs wetting layer (WL) surface), the averaged base diameter D (at the level of $0.1h$), and the surface density N_s.

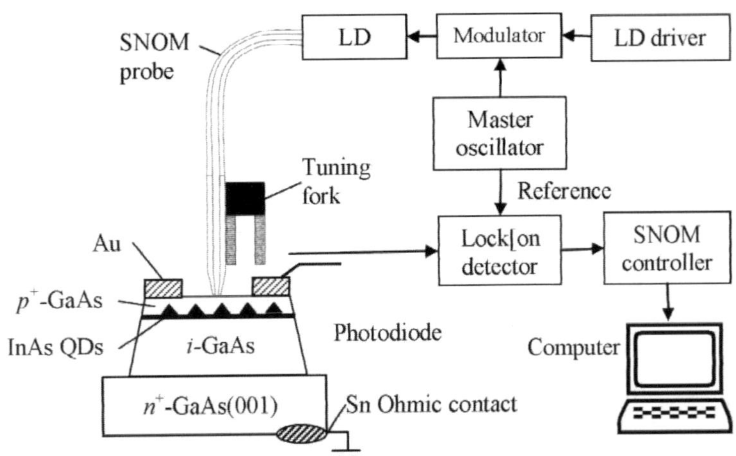

Figure 1. Schematic representation of the experimental setup for the PC SNOM measurements.

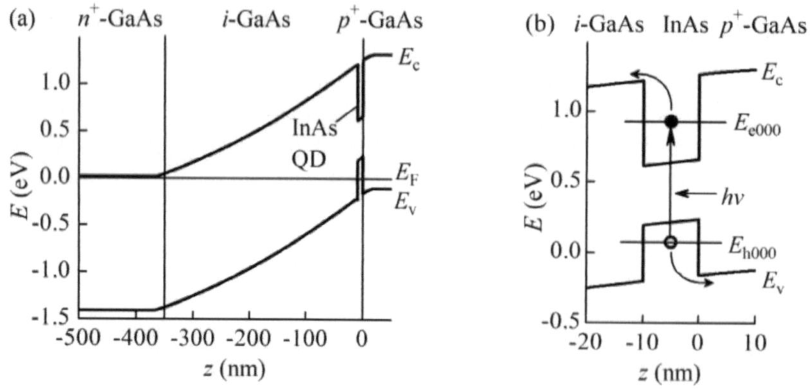

Figure 2. (a) Calculated equilibrium band diagram (300 K) of the GaAs-based *p–i–n* structure with embedded InAs/GaAs(001) QDs for the PC SNOM investigations; (b) the mechanism of PC due to the interband optical absorption in the InAs QDs.

The optical properties of the *p–i–n* diode structure were examined by the photoluminescence (PL) spectroscopy at 77 and 300 K using LOMO® MDR-23 grating monochromator. The PL was excited by a continuous wave (CW) He–Ne laser with $\lambda_{ex} \approx 632.8$ nm and the output power $P \approx 30$ mW and registered by the lock-on technique.

On the base of the *p–i–n* structure, the cylindrical mesa PDs with the input windows in the upper Au Ohmic contacts were fabricated. The mesa diameter was ≈ 350 μm, the window diameter was ≈ 200 μm. The bottom Ohmic contact to the back side of the n^+-GaAs substrate was alloyed from a Sn foil by a spark discharge. The photoelectric properties of the PDs were examined by the PC spectroscopy at 300 K using LOMO® MDR-2 grating monochromator with a 100-W halogen lamp as the photoexcitation source. The spectra of photosensitivity (PS) S_{ph} were measured in the shortcut circuit (SC) photocurrent mode with the chopped photoexcitation (the chopper frequency was ≈ 180 Hz) using Stanford Research® SR-530 lock-in detector and were normalized to the photoexcitation intensity spectrum.

The PC SNOM investigations were carried out in ambient air at 300 K using NT-MDT® Solver™ SNOM scanning head and controller with modulated photoexcitation. The schematic representation of the experimental setup is shown in Figure 1. Dilas® DMPO131-14 and DMPO155-14 single mode CW pigtail semiconductor laser diodes (LDs) with the emission wavelengths $\lambda_{ex} \approx 1310$ nm and ≈ 980 nm, respectively were employed as the photoexcitation sources. Both values of λ_{ex} were within the interband optical absorption band of the InAs QDs in the investigated PDs as follows from the PL and PC spectra (see Section 3 below). The nominal values of *P* in the pigtails were ≈ 1.5 mW for both LDs. However, in the experiments *P* was limited to 200 μW in order to avoid the damage of the SNOM probe aperture due to burning. The power supply for the LDs was provided by Dilas® PILOT™ 4-DC LD driver. The photoexcitation intensity was modulated by modulating the LD pumping current by an optron driven by a master oscillator with the frequency of ≈ 360 Hz. The photoresponse of the PD was recorded in the SC photocurrent mode by the same SR-530 lock-in detector connected to the user input of Solver™ SNOM controller. The NT MDT® SNOM probes with the integrated tuning fork shear force sensors were used. The nominal aperture size was ≈ 100 nm.

3. Results and discussion

An AFM image of the satellite sample with the surface InAs/GaAs(001) QDs is presented in Figure 3a. The histograms of distribution of the QDs in the AFM image by *D* and *h* are presented in Figures 3b and 3c, respectively. The surface QDs observed in the AFM image has a three-modal distribution both by *D* and by *h*. Besides the coherent InAs QDs with *h* = 4-9 nm and *D* = 10-20 nm, the relaxed InGaAs clusters with *D* = 60-120 nm and *h* = 20-40 nm have been observed. The latter arise as a result of coalescence of the coherent InAs QDs during growth [5]. In turn, the coherent InAs QDs have a bimodal distribution by D and h typical for the InAs/GaAs(001) QDs grown by MOVPE [6]. The smaller QDs have D = 8-12 nm and h = 4-5 nm, while the bigger ones have D = 16-20 nm and h = 6-9 nm. It is worth noting that the lateral sizes of the InAs QDs in Figure 3 were enlarged as compared to the actual ones up to 40-50 nm due to the convolution effect [7].

The PL spectra of the *p–i–n* structure with embedded InAs QDs at 77 and 300 K are presented in Figure 4a. Besides the peaks of GaAs edge PL at the photon energies $h\nu \approx 1.51$ and 1.42 eV at 77 and 300 K, respectively, the peaks at $h\nu \approx 1.07$ eV and ≈ 1.25 eV related to the interband radiative optical transitions between the ground quantum confined electron and hole states in InAs QDs of bigger and smaller sizes, respectively were observed in the PL spectrum at 77 K. In the PL spectrum measured at 300 K, the PL peak from the bigger QDs (at $h\nu \approx 1.01$ eV) was observed only because the energy barrier for the electrons occupying the ground state in the QDs appears to be too small (≈ 130 meV according to [4]). It is comparable to the thermal energy $kT \approx 26$ meV at 300 K, so that the thermal emission of the excess electrons from the QDs prevails the radiative recombination inside the QDs [8]. As a result, the PL peak from the smaller QDs was not observed at 300 K.

In the PS spectrum (300 K) shown in Figure 4b, besides the PS band with the edges at $h\nu \approx 1.42$ eV and ≈ 1.35 eV related to the GaAs intrinsic PS edge and to the InAs WL, respectively, the PS bands with the edges at $h\nu \approx 1.2$ eV and ≈ 0.95 eV attributed to the PS of the InAs QDs of the smaller and the bigger sizes, respectively were observed. The PS of the QDs was related to the interband optical transitions in the QDs followed by the emission of the photoexcited carriers from the QDs into the GaAs host matrix and the separation of the electron-hole pairs by the electric field of the *p–i–n* junction, as shown in Figure 2b.

The PS edge at $h\nu \approx 0.95$ eV was attributed to the relaxed InGaAs clusters. One can estimate the In molar fraction x in the cluster material ($In_xGa_{1-x}As$ alloy) from the spectral position of the respective PS band edge as $x \approx 0.35$ [6]. It is worth noting that the PS bands of the bigger QDs and the InGaAs clusters in Figure 4b are ~ 100 times weaker than the one of the smaller InAs QDs. One of the reasons for this is a lower surface density of the bigger QDs than of the smaller ones ($\approx 2.2\times10^8$ cm^{-2} and $\approx 4.9\times10^8$ cm^{-2}, respectively). Another reason is the recombination of the photoexcited carriers inside the bigger QDs and InGaAs clusters prevailing over the emission from these ones. The observation of the PL from the bigger QDs at 300 K supports the above explanation.

Note that the PL lines from the InGaAs clusters are not manifested in the PL spectra, because the relaxed clusters contain the misfit dislocations, so that the non-radiative recombination prevails over the radiative one.

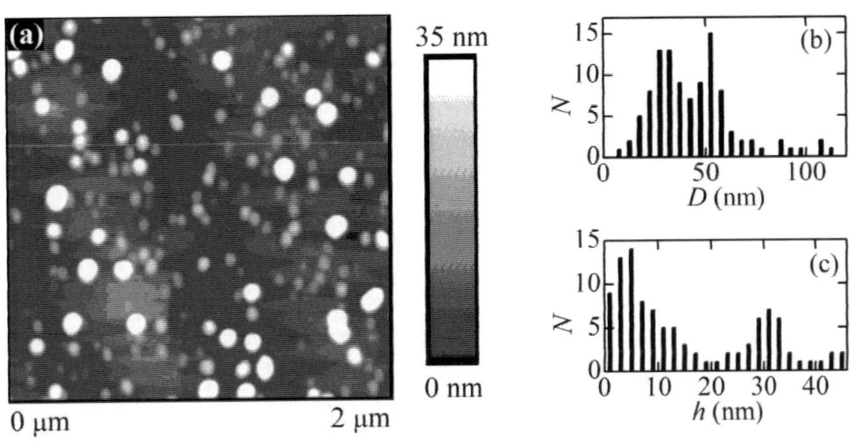

Figure 3. (a) AFM image and the histograms of distribution of the InAs QDs and relaxed InGaAs nanoclusters (b) by base size D and (c) by height h.

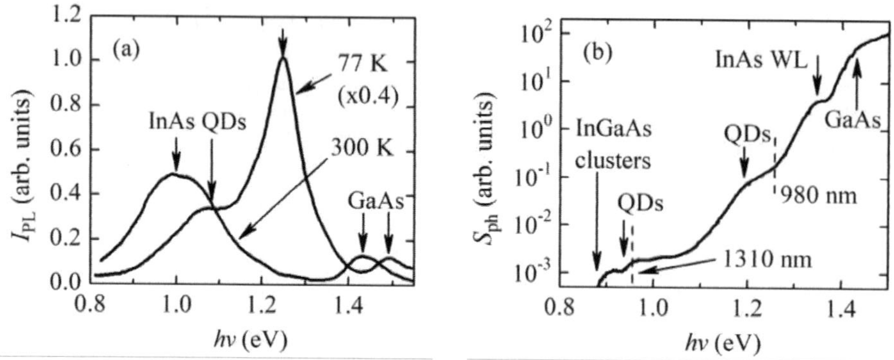

Figure 4. (a) PL spectra (77 and 300 K) of the *p–i–n* structure with InAs QDs; (b) PS spectrum (300 K) of the PD based on the *p–i–n* structure. The dashed lines indicate the photon energies of the emission of the LDs employed in the PC SNOM measurements.

The PC SNOM images recorded at $\lambda_{ex} \approx 1310$ nm and ≈ 980 nm are presented in Figures 5a and 5b, respectively. The inhomogeneities observed in the PC SNOM images were related to the interband optical transitions between the quantum confined electron and hole sates in the InAs QDs as well as between the valence band states and the conduction band ones in the relaxed InGaAs nanoclusters. As it follows from the PC spectrum of the diode shown in Figure 4b, the biggest InAs QDs only as well as the InGaAs clusters can be manifested in the PC SNOM image in Figure 5a recorded at $\lambda_{ex} \approx 1310$ nm. In contrary, the smaller QDs can be manifested in the PC SNOM image recorded at $\lambda_{ex} \approx 980$ nm only. The minimum size of the features observed in the PC SNOM images presented in Figure 5 was ~ 100 nm that coincides to the nominal aperture diameter *a* of the SNOM probes used, which defines the spatial resolution of the method. Since the surface density of the smaller QDs was more than 2 times greater than the one of the bigger QDs, the images of QDs in Figure 5b tend to overlap. It is worth noting that not all QDs are manifested in the PC SNOM images, but only these ones, in which the ground state transition energy (or the one of another allowed interband optical transition between the quantum confined electron and hole states) matches the photoexcitation photon energy. The self-assembled InAs/GaAs(001) QDs are featured by a considerable natural scatter in size and, correspondingly, in the interband transition energies [9]. The width of the latter is much greater than the pumping LD emission linewidth (~ 10 meV). As a result, only a small fraction of QDs falls into the LD emission line energy interval and are manifested in the PC SNOM images.

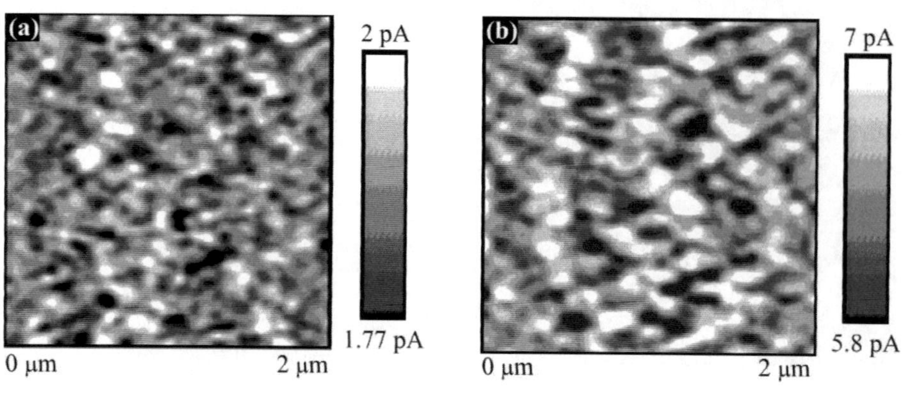

Figure 5. PC SNOM images (300 K) of the input window of a PD based on the *p–i–n* structure with embedded InAs/GaAs(001) QDs. λ_{ex}, nm: (a) 1310; (b) 980.

The rest QDs are manifested in the PC SNOM image as the dark spots on more or less constant background (originating from the impurity PS of the *i*-GaAs layer and of the n^+-GaAs substrate as well as from the stray photoexcitation reflected back from the back side of the substrate) because of the scattering of the photoexcitation outgoing from the SNOM probe aperture on the individual InAs QDs [10]. This is a principal difference of the InAs QDs from the self assembled GeSi/Si(001) nanoislands studied in Ref. [3]. The latter are the type-II heterostructures, so that the initial hole states of the interband optical transitions are confined in the GeSi nanoislands, whereas the final states are the continuous ones in the conduction band of the host matrix material (Si). As a result, all GeSi nanoislands with the ground state interband transition energy E_0, being smaller than the photoexcitation photon energy, are manifested in the PC SNOM images whether the hole states in the GeSi nanoislands are the quantum confined ones or not [3]. In the relaxed InGaAs clusters, the size quantization effect is absent due to relatively large size of these ones as compared to the de Broglie wavelength of the electrons and holes. As a result, all relaxed InGaAs clusters are manifested in the PC SNOM images in Figure 5 as the white spots, which sizes and surface density ($\approx 4\times10^8$ cm^{-2}) coincide with the ones of the relaxed InGaAs clusters in the AFM image in Figure 2a.

4. Conclusion

The results of the present study demonstrate the possibility of observation of the photoresponse from single InAs/GaAs(001) QD due to the interband optical transitions between the quantum confined electron and hole states in the QD at the photoexcitation through a SNOM probe and of imaging the individual InAs/GaAs(001) QDs in the PC SNOM images. The spatial resolution of PC SNOM was limited by the SNOM probe aperture diameter ~ 100 nm. Not all QDs are manifested in the PC SNOM images but only these ones, in which the energy of the ground state interband optical transition (or another allowed interband transition) between the quantum confined electron and hole states match the photoexcitation photon energy. The rest QDs are manifested as the dark spots in the PC SNOM images due to the scattering of the light incident from the SNOM probe by the QDs.

Acknowledgements

The work was supported by RFBR (16-02-00450). The SNOM measurements have been carried out using the shared research facilities of Research and Education Center for Physics of Solid State Nanostructures at Lobachevskii State University of Nizhnii Novgorod.

References

[1] Paesler M A and Moyer P J 1996 *Near-Field Optics: Theory, Instrumentation, and Applications* (NY, Wiley Interscience)

[2] Cricenti A, Generosi R, Barchesi C, Luce M and Rinaldi M 1998 *Rev. Sci. Instrum.* **69** 3240

[3] Filatov D O, Kazantseva I A, Shengurov V G, Chalkov V Yu, Denisov S A, Gorshkov A P and Mishkin V P 2017 *Semicond.* **51** 536

[4] Stier O, Grundmann M and Bimberg D 1999 *Phys. Rev.* B **59** 5688

[5] Karpovich I A, Zvonkov B N, Baidus' N V, Tikhov S V and Filatov D O 2004 *Trends in Nanotechnology Research* (NY, Nova Science), 173

[6] Karpovich I A, Baidus' N V, Zvonkov B N, Morozov S V, Filatov D O and Zdoroveishev A V 2001 *Nanotechnology* **12** 425

[7] Bukharaev A A, Berdunov N V, Ovchinnikov D V and Salikhov K M 1998 *Scanning Microscopy* **12** 225

[8] Volkova N S, Gorshkov A P, Filatov D O and Abramkin D S 2014 *JETP Lett.* **100** 175

[9] Ledentsov N N, Ustinov V M, Shchukin V A, Kop`ev P S, Alferov Zh I and Bimberg D 1998 *Semicond.* **32** 385

[10] Karpovich I A, Filatov D O, Morozov S V, Baidus N V, Zvonkov B N and Gushina Yu Yu 2000 *Bulletin RAS: Physics* **63** 313

IOP Conf. Series: Materials Science and Engineering **256** (2017) 012006 doi:10.1088/1757-899X/256/1/012006

The generation of the electrical oscillations in a contact of an AFM probe to an individual Au nanoparticle in a SiO_2/Si film

D O Filatov[1,*], O N Gorshkov[1,2], D A Antonov[1], M E Shenina[1], D Yu Sinutkin[3], A V Zenkevich[4] and Yu A Matveev[4]

[1] Research and Education Center for Physics of Solid State Nanostructures, Lobachevskii State University of Nizhnii Novgorod, Nizhnii Novgorod, 603950, Russia
[2] Research Institute for Physics and Technology, Lobachevskii State University of Nizhnii Novgorod, Nizhnii Novgorod, 603950, Russia
[3] Department of Physics, Lobachevskii State University of Nizhnii Novgorod, Nizhnii Novgorod, 603950, Russia
[4] Moscow Institute of Physics and Technology, Moscow, 141701, Russia

*dmitry_filatov@inbox.ru

Abstract. We report on the experimental observation of the electrical oscillations in an oscillating loop connected to a contact of a conductive probe of an atomic force microscope to a tunnel-transparent (≈ 5 nm thick) SiO_2/Si(001) film with embedded Au nanoparticles. The oscillations were attributed to the negative differential resistance of the probe-to-sample contact originating from the resonant electron tunnelling between the probe and the Si substrate via the quantum confined electron energy states in small (~1 nm in size) Au nanoparticles. This observation demonstrates the prospects to build an oscillator nanoelectronic device based on an individual nanometer-sized metal nanoparticle.

1. Introduction

Metal nanoparticles (NPs) embedded into dielectric films have attracted much attention in recent years due to a variety of potential applications in non-volatile memory, plasmonics, nonlinear optics, etc. [1]. Earlier, we have demonstrated the current-voltage (*I–V*) curves of a contact between a conductive probe of an atomic force microscope (AFM) and a tunnel transparent (3 to 5 nm thick) SiO_2/Si(001) film with embedded Au NPs deposited by Pulsed Laser Deposition (PLD) to manifest the negative differential resistance (NDR) attributed to the resonant electron tunneling via the quantum confined electron states in small enough (~1 nm in thickness) Au NPs [2]. It is well known that once a nonlinear electrical element with NDR is available, one can build an oscillator by combining this NDR element with an oscillating loop. In Ref. [3], we have reported an experimental observation of the electrical oscillations in an oscillating loop connected to a contact of a conductive AFM (CAFM) probe to a tunnel-transparent yttria stabilized zirconia (YSZ) film with embedded Au NPs. The excitation of the electrical oscillations was attributed to NDR of the CAFM probe-to-sample contact originating from the resonant electron tunneling via the quantum confined electron states in small Au NPs. This result demonstrated experimentally the prospects to build an oscillator nanoelectronic device based on an individual metal NP.

In the present paper, we report on the experimental observation of the electrical oscillations in the oscillating loop connected to the CAFM probe contact to SiO_2/Si(001) film with embedded Au NPs.

2. Experimental details

The SiO_2 films with embedded Au NPs were fabricated using self-assembling during the plasma-enhanced oxidation [4] of the amorphous Au–Si mixture (1:6) films of ≈1.5 nm in thickness deposited by PLD at 300 K onto the n^+-Si(001) substrates covered by the native oxide. The Au NPs of 5–7 nm in the lateral size, according to Transmission Electron Microscopy (TEM) data [4], and of 1–1.5 nm in thickness were arranged almost in single sheet inside ≈3.5 nm thick SiO_2/Si film (Fig. 1).

The CAFM investigations were carried out in Ultra High Vacuum (UHV) environment (the base residual gas pressure was $<10^{-10}$ Torr) at 300 K using Omicron® UHV AFM/STM LF1 atomic force/scanning tunneling microscope (AFM/STM) installed into Omicron® MultiProbe™ RM UHV system. The schematic representation of the CAFM experiment is shown in Figure 1. The sample surface was scanned across in Contact Mode by a conductive AFM probe (NT-MDT® NSG-11 DLC diamond-like film coated probes were used). A bias voltage V_g was applied to the n^+-Si substrate relative to the AFM probe and the electric current between the probe and the substrate I_t was measured as a function of the probe coordinates in the sample surface plane (x, y). In another measurement mode, the I–V curves of the probe-to-sample contact were measured in every point of the AFM scan.

In order to observe the oscillations of I_t due to NDR of the probe-to-sample contact, an oscillating loop consisting of a capacitor C and a coil L was connected between the sample and the bias voltage source (Fig. 1). The free oscillation frequency of the oscillating loop f_0 was chosen to be ≈24.3 kHz, because the bandpass of the STM preamplifier (I–V converter) was limited to 30 kHz at the factory in order to suppress the high frequency noise. The waveforms of the probe current $I_t(t)$, where t is the time, were recorded by Agilent 3000A oscilloscope connected to the STM preamplifier output.

Capturing the waveform $I_t(t)$ at the moment, when the excitation of the oscillations takes place, was the key issue of the experiment. Indeed, the following conditions should be satisfied for the electrical oscillations in the oscillating loop to occur:

- the AFM probe should be positioned just above an Au NP and the NDR should be present in the I–V curve of the probe-to-sample contact;
- the bias voltage V_g should be within the NDR region in the I–V curve.

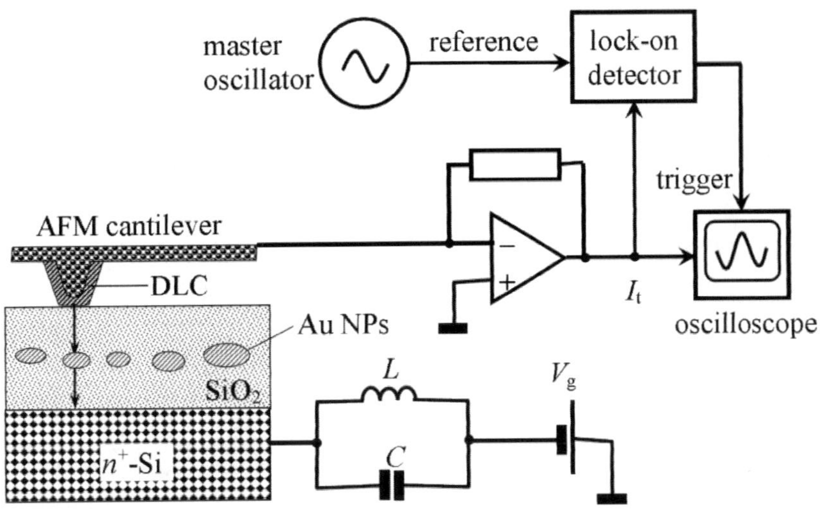

Figure 1. Schematic representation of the CAFM experiment.

These conditions are difficult to satisfy simultaneously because of the creep of the AFM piezo scanner. The AFM probe having been positioned onto a NP drifts away very quickly. Therefore, in order to record the waveforms of the electrical oscillations in the oscillation loop, the procedure described earlier in [3] was applied. The *I–V* curves were recorded in every point of the AFM frame. The limits of the V_g sweep were set from 0 to 8 V in order to ensure the NDR region in the *I–V* curves to be passed through. Stanford Research® SR-530 lock-in detector has been connected to the user's "I_t" output of Omicron® AFM/STM control electronics in parallel to the oscilloscope (Fig. 1). The reference signal for the lock-in detector has been provided by a high precision digital generator serving as a master oscillator. The frequency of the reference signal was set equal to the free oscillation frequency of the oscillating loop f_0. The output of the lock-in detector was connected to the trigger input of the oscilloscope. In the course of scanning, once the AFM probe encounters a NP and V_g reaches the beginning of the NDR region, the excitation of the electrical oscillations begins in the oscillating loop and the sinusoidal component with the frequency f_0 appears in the spectrum of the probe current I_t. Hence, the output signal of the lock-in detector increases with increasing oscillations' amplitude at the frequency f_0. Once the output signal of the lock-in detector reaches a preset threshold of the oscilloscope trigger action, the waveform $I_t(t)$ is captured.

3. Results and discussion

An AFM image and current one, i. e. a map of $I_t(x, y)$ at given V_g of a SiO_2:NP-Au/Si film are shown in Figures 2a and 2b, respectively. The spots of increased I_t (hereafter referred to as the current channels) in Figure 2b were related to the electron tunneling between the AFM probe coating and the n^+-Si substrate via the Au NPs [2] as shown in Figure 1.

The sizes of the current channels in Figure 2b ranged from ≈30 to ≈100 nm, whereas the lateral sizes of the Au NPs were 5 to 7 nm, according to the TEM data [4], as it has been already mentioned above. This disagreement can be explained as follows. In general, the sizes of the CAFM images of the metal NPs inside the dielectric films are determined by the diameter of the contact area between the AFM probe tip and the sample surface D_p and do not depend on the size of the NPs themselves [5]. In turn, D_p depends on the tip curvature radius R_p and on the loading force, as well as on the elastic modules of the sample and probe coating materials. Roughly, one can estimate $D_p \sim R_p \approx 70$ nm for NT-MDT® NSG-11 DLC probes (according to the vendor's specifications) that is consistent with the sizes of the current channels in Figure 2b. *I–V* curve of the contact between the AFM probe and the SiO_2:NP-Au/n^+-Si(001) film measured inside a current channel is shown in Figure 3. The current channel, wherein the *I–V* curve was measured, is marked by a circle in Figure 2b. A well expressed NDR at the positive bias V_g applied to the n^+-Si substrate relative to the AFM probe (corresponding to the injection of the electrons from the AFM probe into the SiO_2:NP-Au/Si film) has been observed.

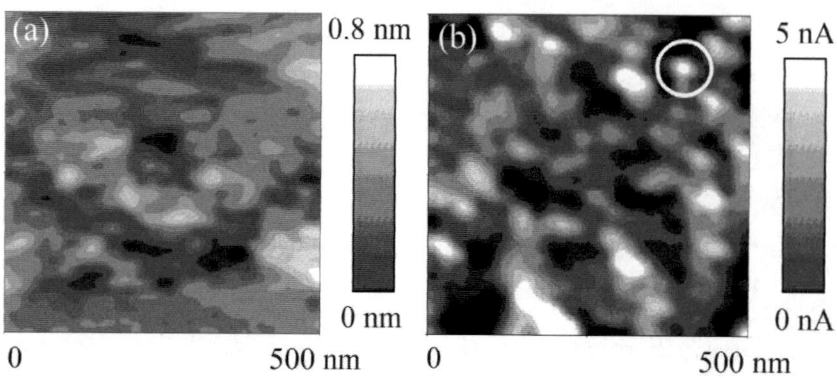

Figure 2. (a) AFM and (b) current images of the SiO_2:NP-Au/n^+\$-Si(001) film. $V_g = 4.5$ V.

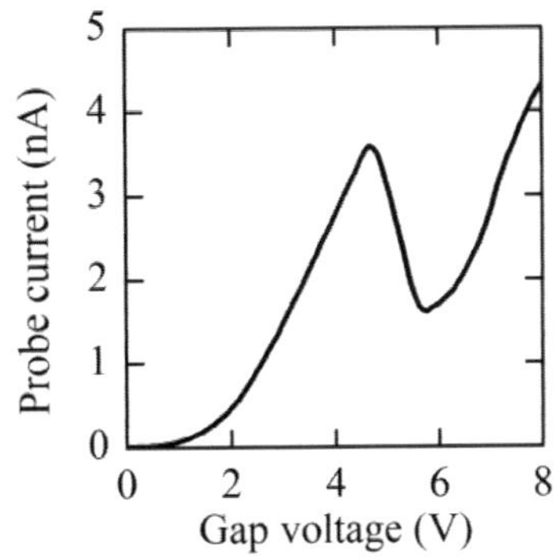

Figure 3. *I–V* curve of the probe-to-sample contact measured in the current channel marked by a circle in Figure 2b.

The NDR was attributed to the resonant electron tunneling from the AFM probe coating into the n^+-Si substrate via the quantum confined electron states in the Au NPs [2]. A one-dimensional band diagram of the contact between the DLC-coated AFM probe and the SiO_2:NP-Au/n^+-Si film across an Au NP of 1 nm in thickness embedded into the middle of a 3.5 nm thick SiO_2 film calculated for $V_g = 4.5$ V is presented in Figure 4a. The calculation procedure was described in details in Ref. [2]. One can see in Figure 4a that the SiO_2/Au/SiO_2/Si stack represents a double barrier structure (DBS) for the electrons similar to the one of the resonant tunnel diode [6].

Figure 4b presents the calculated tunnel transparency spectrum T(E) (300 K) of the DBS presented in Figure 4a for $V_g = 4.5$ V, which resonant peak in the *I–V* curve has been observed in Figure 3. The procedure of the calculation was described in Ref. [2] as well. The peaks in T(E) correspond to the virtual quantum confined energy states in the Au NP. For example, the calculated envelope wavefunction $\chi(z)$ for the resonant state marked by the arrow in Figure 4b is shown in Figure 4a. The energy spacing between the resonant states in the Au NPs can be estimated as ≈ 1.5 eV that is much greater than the thermal energy kT at 300 K (≈ 26 meV). So far, the conditions for the manifestation of the resonant tunneling effect were readily available in the investigated films.

In Figure 5, a probe current waveform $I_t(t)$ recorded at $V_g \approx 5$ V is presented, i. e. V_g was within the NDR region of the *I–V* curve of the probe-to-sample contact (Fig. 4a). Nearly sinusoidal oscillations of I_t have been observed. The frequency of the oscillations matches the free oscillation frequency of the oscillating loop $f_0 \approx 24.3$ kHz.

Thus, the results of the present study demonstrate the possibility to build an oscillator based on the resonant tunneling effect in a single Au NP embedded into a tunnel transparent SiO_2/Si film.

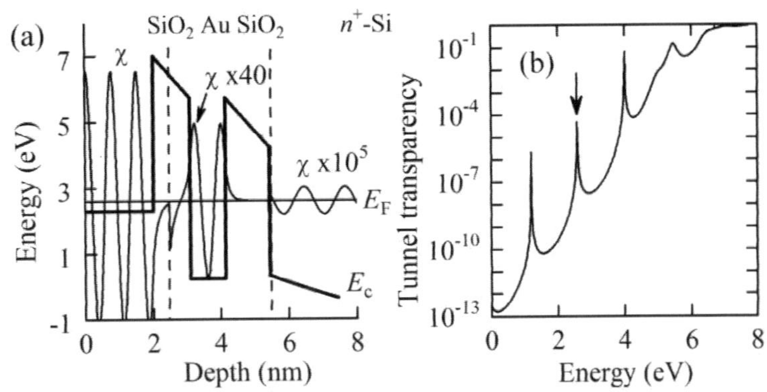

Figure 4. (a) Calculated band diagram and (b) tunnel transparency spectrum of a contact between the AFM probe and the SiO_2:NP-Au/n^+-Si film. $V_g = 4.5$ V.

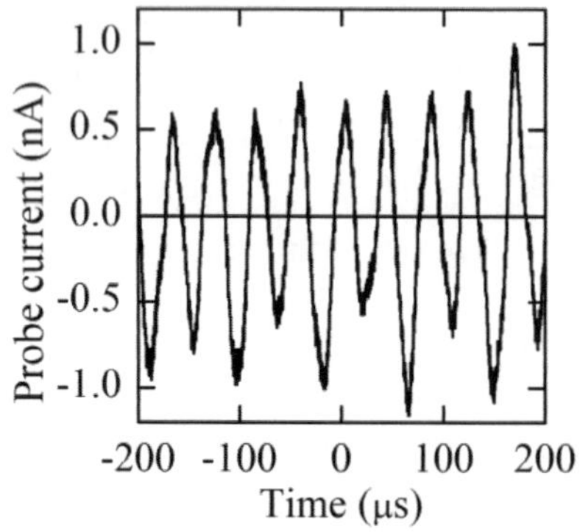

Figure 5. A waveform of the probe current oscillations.

Future prospects of the device application of these results are connected with the replacement of the AFM probe by a permanent point electrical contact to individual metal NPs. The present microelectronics is approaching the 10-nm critical dimension rapidly, so one can expect an electrical contact of ~ 10 nm in size with appropriate microstrip wiring to become possible in the nearest future.

It is worth noting that similar results obtained in Ref. [3] on an YSZ:NP-Au film and on the SiO_2:NP-Au ones in the present study point to the origin of the NDR in the *I–V* curves to be not the properties of the film materials, but the fundamental phenomenon of resonant tunneling via the ultrafine Au NPs.

It should be noted also that a low free oscillation frequency of the oscillation loop f_0 was selected in the present work for the reason of limited bandpass of the STM preamplifier only. Since the tunneling process itself is very fast, the operation frequency of the nanoelectronic oscillator device based on a single metal NP could be raised, in general, up to the THz band when an appropriate cavity is utilized.

4. Conclusion

In the present study, we have demonstrated experimentally the room-temperature operation of a prototype nanoelectronic oscillator device based on a single Au NP embedded into a tunnel-transparent SiO_2/Si film. The operation of the device is based on the resonant electron tunneling via the quantum confined energy states in an ultrafine Au NP. To provide a local tunnel electrical contact to a single NP inside the SiO_2 film, we applied a probe of an atomic force microscope. The results of the present study demonstrate wide prospects for future application of such nanoelectronic devices.

Acknowledgements

This work has been supported by Ministry of Education and Science of Russian Federation (Project 16.7864.2017). The CAFM measurements have been carried out using the shared research facilities of Research and Education Center for Physics of Solid State Nanostructures at Lobachevskii State University of Nizhnii Novgorod.

References

[1] Lue J-T 2001 *J. Phys. Chem. Solids* **62**, 1599
[2] Filatov D O, Lapshina M A, Antonov D A, Gorshkov O N, Zenkevich A V and Lebedinskii Yu Yu 2010 *J. Phys.: Conference Series* **245** 012018
[3] Gorshkov O N, Filatov D O, Antonov D A, Antonov I N, Shenina M E and Pavlov D A 2017 *J. Appl. Phys.* **121** 014308
[4] Zenkevich A V, Lebedinskii Yu Yu, Timofeyev A A, Isayev I A and Tronin V N 2009 *Appl. Surf. Sci.* **255** 5355
[5] Lapshina M A, Filatov D O and Antonov D A 2008 *J. Surf. Invesigation: X-ray, Synchrotron and Neutron Techn.* **2** 616
[6] Chang L L, Esaki L and Tsu R 1974 *Appl. Phys. Lett.* **24** 5935

IOP Conf. Series: Materials Science and Engineering **256** (2017) 012007 doi:10.1088/1757-899X/256/1/012007

Application of focused ion beam for the fabrication of AFM probes

A S Kolomiytsev*, S A Lisitsyn, V A Smirnov, A A Fedotov and Yu N Varzarev

Southern Federal University, Institute of Nanotechnologies, Electronics, and Electronic Equipment Engineering, Taganrog, 347922, Russia

*askolomiytsev@sfedu.ru

Abstract. The results of an experimental study of the probe tips fabrication for critical-dimension atomic force microscopy (CD-AFM) using the focused ion beam (FIB) induced deposition are presented. Methods of the FIB-induced deposition of tungsten and carbon onto the tip of an AFM probe are studied. Based on the results obtained in the study, probes for the CD-AFM technique with a tip height about 1 μm and radius of 20 nm were created. The formation of CD-AFM probes by FIB-induced deposition allows creating a high efficiency tool for nanotechnology and nanodiagnostics. The use of modified cantilevers allows minimizing the artefacts of AFM images and increasing the accuracy of the relief measurement. The obtained results can be used for fabrication of AFM probes for express monitoring of the technological process in the manufacturing of the elements for micro- and nanoelectronics.

1. Introduction

An important stage in the development and production of integrated circuits is the implementation of operational control of geometric parameters of structures at various stages of the technological process. With modern topological norms of the production of very-large-scale integration (VLSI) devices (less than 30 nm), the microrelief of the surface both in the horizontal and in the vertical planes often plays a decisive role in obtaining a positive result [1]. Thus, an important task is the development of procedures and tools for determining the geometric parameters and the roughness of the vertical walls in high-aspect nanostructures.

At present, atomic force microscopy (AFM) is one of the most promising methods for studying the structure and properties of the solids surface with nanometer spatial resolution [2, 3]. Critical-dimension atomic-force microscopy (CD-AFM) is a promising method that can effectively solve problems of measuring the roughness of vertical walls by examining the morphology and local surface properties of high-aspect structures with a high spatial resolution [4-7]. The resolving power of an AFM is determined by the shape and radius of curvature of the used probe, which are limited by the technological processes, which entails the appearance of artifacts in AFM studies [8]. CD-AFM is based on the principle of detecting the force interaction of the lateral surface of the probe tip with the vertical planes of the structure under study (Fig. 1). For this purpose, protrusions of about 20 nm in length are axially symmetrically formed on the lateral surface of the probe.

The main problem of the CD-AFM method is the complexity and high cost of the probes manufacturing. Using local methods of nanoscale structuring, it is possible to create tips of CD-AFM probes with improved parameters, which allow minimizing artifacts, reducing costs, and increasing the accuracy and resolution of microscopy. One of the promising methods for forming the CD-AFM probe

IOP Conf. Series: Materials Science and Engineering **256** (2017) 012007 doi:10.1088/1757-899X/256/1/012007

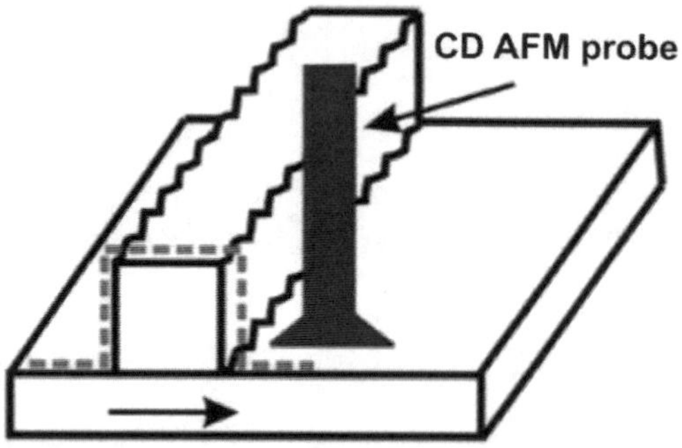

Figure 1. Principle of CD-AFM mode.

tips is the use of focused ion beam (FIB) [9-11]. The essence of the FIB method lies in the local ion-beam sputtering of the material by a flow of gallium ions focused to a diameter of about 7 nm. The combination of the FIB and highly selective gas chemistry allows realizing operations of local ion-induced deposition of various materials (carbon, tungsten, platinum, etc.). Precision ion beam milling and ion-induced deposition makes it possible to form the tips of AFM probes with unique parameters.

The goal of our study is to examine methods for the modification and reconditioning of AFM probes via material deposition onto a tip using FIB-induced deposition to form CD-AFM probes.

2. Experimental

The fabrication of the probe tips was performed with a FEI Company DualBeam system Nova NanoLab 600, combining a Ga^+ FIB and a field emission scanning electron microscope. AFM experiments were performed using a Ntegra Vita AFM system by NT-MDT.

At the initial stage, the experimental studies of the ion-induced deposition of carbon and tungsten on the silicon surface were carried out. It was found that for accelerating voltage of 30 keV and ion beam current of 1 nA, the deposition rate of carbon was 7.8 nm/sec and of tungsten – 3.5 nm/sec. The minimum diameters of the formed structures were 22 nm and 35 nm, respectively.

The tips of the CD-AFM probes were formed on the basis of the standard AFM cantilevers NSG-10 (resonant frequency: 290 kHz, force constant: 44 N/m) with the damaged tip after intensive use. At the first stage of the experiment, the old tip of the probe was removed by FIB milling, and a basis for building up the new tip was formed. The following FIB parameters were used: the accelerating voltage of the ion beam – 30 keV; the ion beam current – 1 nA; and the dwell time of the ion beam – 1.0 μs. After that, the probe tip was formed by the method of FIB-induced carbon deposition. A probe with a tip height of about 1.2 μm and a diameter of about 150 nm was formed. The following parameters of ion-stimulated deposition of carbon were used: the ion beam current – 0.1 nA, the time of action at the point – 100 μs, the accelerating voltage – 30 keV. The chamber pressure after introducing $C_{14}H_{10}$ gas was 1×10^{-4} Pa. A bitmap of the desired probe structure was created using Unigen 3.2 software, and then uploaded into the FIB software [9]. At the next stage of research, rounded protrusions of about 120 nm in length and a radius of 20 nm were formed perpendicularly to the probe tip (Fig. 2). With these parameters, the process of a probe tip formation takes about 5 min, including the milling and deposition operations.

We tested a performance, resolution, and reliability of the FIB-fabricated probe by taking CD-AFM images of the 500 nm step of PMMA photoresist on the silicon surface. The AFM study of the lateral surface roughness was carried out at the Ntegra Vita probe laboratory using a standard NSG-10 probe and a FIB-fabricated CD-AFM probe. A semicontact AFM image of the PMMA step was obtained with the FIB-fabricated probe, demonstrating its capability for high spatial resolution imaging of the sidewalls. The resonance frequency of the new cantilever was 281,900 Hz, being not much different from that of the commercial cantilever. Figure 3 shows the AFM images of the PMMA step surface and profile obtained by FIB-fabricated probe.

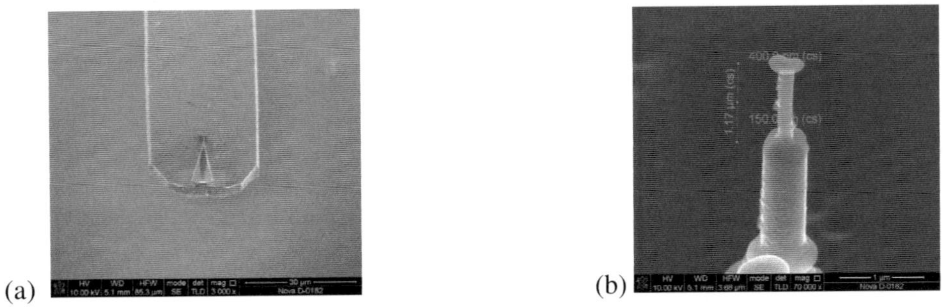

Figure 2. SEM-image of the FIB-fabricated CD AFM probes.

3. Results and discussion

The analysis of AFM images shows that the sidewall roughness of the PMMA structures, obtained using standard cantilevers, contains artefacts and cannot be analyzed and measured. The artefacts were absent (Fig. 3) on the AFM image of the PMMA step scanned by the FIB-modified cantilever. As shown in Figure 3c, the use of a modified probe allows measuring the relief and roughness of the PMMA sidewall. The software package Image Analysis 3.0 (NT-MDT Co) allowed measuring the arithmetic mean roughness of the relief of the lateral wall, which was 9,44±1,96 nm.

It was also found that in order to achieve a high rate of ion-induced deposition the high current values (0.1-1 nA) should be used, but in this case it is impossible to achieve high resolution of deposition. To form structures with dimensions less than 50 nm, it is recommended to use small currents of the focused ion beam (30-50 pA).

The presented technique allows restoring the functionality of worn-out and broken AFM probes and improving their properties and parameters. Restoring the functionality of broken probes by FIB is cost-effective, because the standard probes are not maintainable. To automate the process of AFM probe fabrication by FIB, Unigen 3.2 software is used, which was developed in the Southern Federal University [9]. The software allows to generate patterns for FIB milling and deposition as BMP files or ASCII based stream files based on designed 3D profile (as a mathematical formula, a curve, or a standard figure).

The results clearly demonstrate that the FIB-fabricated probes work well as AFM probes for high resolution sidewall roughness imaging (the study of the surface of the ICs, MEMS devices, etc.).

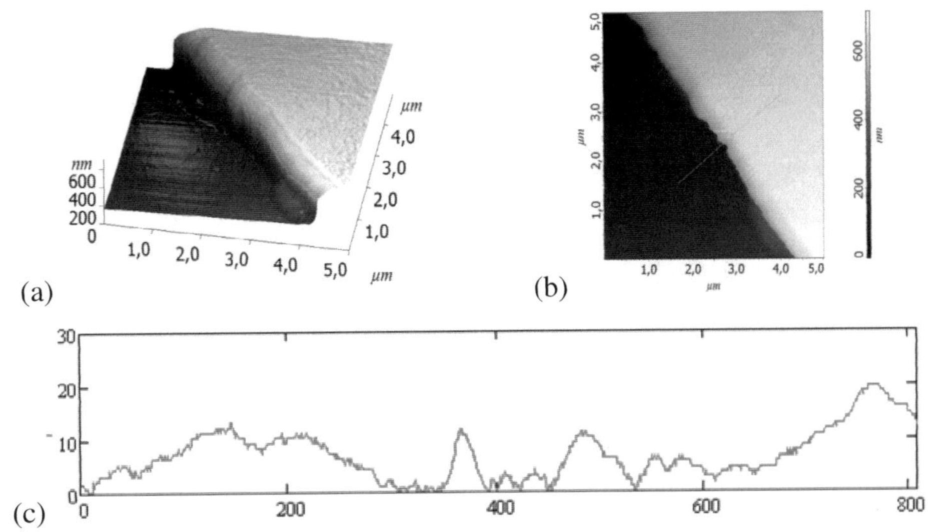

Figure 3. (a, b) AFM images and (c) profile of 500 nm PMMA step obtained using FIB-fabricated CD AFM probe.

4. Conclusion

The experimental study of methods used to modify probes for atomic-force microscopy by the deposition of materials onto the tip of a worn probe using focused ion beam was carried out. Application of the ion-induced deposition allowed creating the probe tips in a wide range of shapes and sizes for different AFM applications. We illustrated the procedure, based on FIB milling and deposition, to create probes with specific tip shape, starting from commercial ones. It was shown that the FIB method provided the formation of the CD-AFM probes with the radius of tip flare rounding 20 nm, tip height 1.2 μm, diameter of about 150 nm, and aspect ratio 1:20. It was shown that the use of modified cantilevers for the diagnostics of sidewall roughness allowed minimizing the artefacts of AFM images, as well as increasing the resolution and reliability of the obtained results.

The obtained results can be used in the development of technological processes for the fabrication and modification of special probes for atomic-force microscopy, including CD-AFM probes, and in the development of procedures for the express monitoring of parameters of the technological process for manufacturing elements for micro- and nanoelectronics and micro- and nanosystems engineering.

Acknowledgments

This work was supported by The Grant of the President of the Russian Federation for State Support of Young Russian Scientists, Candidates of Sciences (project no. MK- 6163.2016.8), with the use of the equipment of the Scientific-Educational Center and the Center for Shared Use "Nanotechnologies" of the Southern Federal University (Taganrog).

References

[1] Bhushan B *2010 Springer Handbook of Nanotechnology (3rd edn)* (New York: Springer) p 1964.
[2] Ageev O A, Konoplev B G and Smirnov V A 2007 *Russ. Microelectron.* **36** 53.
[3] Yao N and Wang Z 2005 *Handbook of Microscopy for Nanotechnology* (Kluwer Acad. Publ.: New York), p 743.
[4] Orji N G, Dixson R G 2007 *Meas. Sci. Technol.* **18** 448.
[5] Ageev O A, Konoplev B G and Smirnov V A 2012 *Russ. Microelectron.* **41** 41.
[6] Ageev O A, Bykov A V, Kolomiitsev A S, Konoplev B G, Rubashkina M V, Smirnov V A and Tsukanova O G 2015 *Semiconductors* **49** 1743–8.
[7] Savenko A, Yildiz I and Petersen D H 2013 *Nanotechnology* **24** 465701.
[8] Ageev O A, Kolomiytsev A S and Bykov A V 2015 *Microelectron. Reliab.* **55** 2131–4.
[9] Ageev O A, Alekseev A M, Vnukova A V, Gromov A L, Kolomiytsev A S, Konoplev B G and Lisitsyn S A 2014 *Nanotechnol. Russia* **9** 26–30.
[10] Giannuzzi L A and Stevie F A 2004 *Introduction to Focused Ion Beams: Instrumentation, Theory, Techniques and Practice* (New York: Springer) 357.
[11] Lisitsyn S A, Kolomiytsev A S, Ilyin O I, Ilyina M V, Konoplev B G, Bykov A V and Ageev O A 2017 *Russ. Microelectron.* **46** 1–6.

Thermal excitation contribution into the electromechanical performance of self-supported Gd-doped ceria membranes

A D Ushakov[1,*,], E Mishuk[2,**], D O Alikin[1,3], B N Slautin[1], A A Esin[1], I S Baturin[1], V Ya Shur[1], I Lubomirsky[2] and A L Kholkin[1,3]**

[1] School of Natural Sciences and Mathematics, Ural Federal University, Ekaterinburg 620000, Russia
[2] Department of Materials and Interfaces, Weizmann Institute of Science, Rehovot, Israel
[3] Department of Physics and CICECO – Aveiro Institute of Materials, University of Aveiro, 3810-193 Aveiro, Portugal

[*]andrey.ushakov@urfu.ru

[**]A. D. Ushakov and E. Mishuk contributed equally to this work

Abstract. Here we present the results of the study of electromechanical performance of the $Ce_{0.8}Gd_{0.2}O_{1.95}$ (CGO) self-supported thin circular membranes with aluminum and titanium as contact electrode materials. The electromechanical performance of both membranes was investigated using highly sensitive interferometric technique and showed two principal excitations mechanisms: common electrostriction and thermal contribution due to the Joule heating. Operating the membranes at frequencies about MHz results in significant contribution of the thermal excitation due to a large power dissipation for every electrode. The excitation of the membrane with aluminum electrodes at the frequencies about 1 Hz leads to the formation of the Schottky barrier at the interface with CGO. That is why the electromechanical response was almost independent on frequency and electric field. Membrane with titanium electrodes showed a prevalence of electrostriction effect with weak frequency dispersion and significant enhancement of the response with temperature.

1. Introduction

The area of microelectromechanical systems (MEMS) is extensively growing during the past forty years. The leading MEMS materials are mainly piezoelectrics in light of their superior sensing and actuating performances: $Pb(Zr,Ti)O_3$ (PZT) and *(1–x)*$PbMg_{1/3}Nb_{2/3}O_3$-$x$$PbTiO_3$ ceramics, thin and thick films [1, 2]. However, the significant electromechanical (EM) response was found only in a few compositions of the morphotropic phase boundary and all of them contain lead, which has negative effect on the human health not only during use, but also during production and recycling [3, 4]. Recently, several multifunctional materials based on the oxygen ionic conductors have been suggested as a substitution of lead containing piezoelectrics, such as (Y, Nb)-stabilized δ-Bi_2O_3 [5], partially hydrated Y-doped barium zirconate $BaZrO_3$, and (Sr,Mg) doped $LaGaO_3$ [5]. Besides, this approach is also attractive due to the absence of hysteresis of the EM response, which limits the creeps and sticking in the final devices [6].

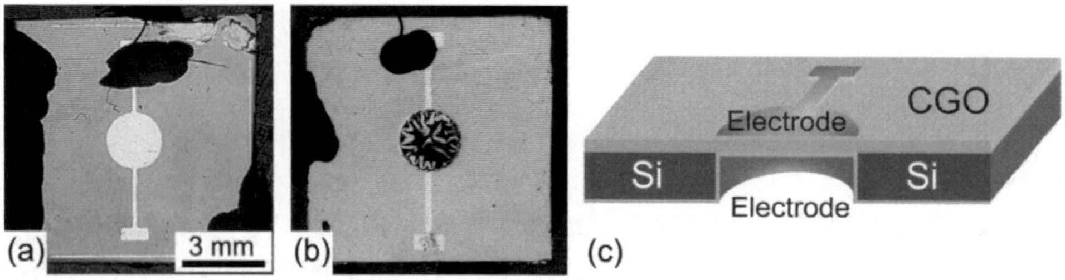

Figure 1. Optical images of CGO self-supported membranes with (a) Ti and (b) Al contacts. (c) Schematic of the membranes.

Gd-doped cerium oxide (CGO) was shown to have a very large electrostriction response [7]. This ionic conductor material is well studied from the point of usage in the solid fuel cells, while its non-typical EM properties were not studied. It was not considered for years as EM material due to its low dielectric constant and expected low electrostriction response according to the Newnham conception [8]. However, recently, the measurement of its EM performance by the cantilever approach revealed the electrostriction coefficient about 3×10^{-16} m^2/V^2 at frequencies below a few Hz [7]. This provides perspectives for its application in MEMS. However, the physical nature of the observed properties has not been completely understood [9].

Here we focus on the problem of thermal excitation appearing as a result of Joule heating that can contribute to the EM performance of CGO self-supported membranes. We have demonstrated that the electrode material [10] becomes even more important in the high operating frequency range (100 Hz-10 kHz). The principally different behavior of the membranes with Al and Ti electrodes at low frequencies is attributed to the formation of Schottky-barriers at the Al electrode/material interface, while at frequency about few MHz both membranes become thermally excited.

2. Experimental

A 1.5-μm-thick layer of $Ce_{0.8}Gd_{0.2}O_{1.9}$ (>99.99% pure, 3-inch ceramics target, 3.175 mm thick) was deposited by RF magnetron sputtering on a Si wafer (intrinsic, 2-inch, 275 μm thick, [100] oriented, double-sided polished, University Wafer, Inc.) covered by a 100 nm thick layer of Al and bottom electrode of Al or Ti (100-nm-thick, >99.999%, Tzamal D-Chem Ltd.). The top Al or Ti electrode, 200-nm-thick, was deposited by magnetron sputtering and patterned lithographically. Annealing at 430 °C for 4 h was carried out under vacuum of about 130 μPa. The cylindrical notch in the Si aligned with the top electrode was fabricated by dry reactive ion etching using back-side patterning of a 1 μm thick Al film that served as a mask [9]. Optical images and schematics of the membrane are shown in Figure 1. The top contact was connected to a copper wire and the bottom contact was connected to the metal pad using silver paint.

Highly sensitive (resolution about 0.1 pm) Michelson interferometer with phase locking and stabilizing PID-feedback loop was employed for the detection of vertical displacements in the middle of membrane. More details of the experimental setup were discussed in [11]. Dielectric measurements were done with QuadTech 7600 Plus (IET Labs, Inc.) RLC-meter.

3. Results and discussion

It is well known that electrostriction is an intrinsic property of all dielectrics whether they are polar or not. It originates from the interaction of the external electric field with the polarization of the unit cell induced by this field. Therefore, the electrostriction response is quadratic with respect to the applied electric field and, if the applied field is sinusoidal, the response could be detected at the second harmonic of excitation [12]. We have found that the EM response of the membranes with Al electrodes in the frequency range from 2 Hz to 1 kHz is much lower than that of the same membranes with Ti electrodes. In the same experimental conditions, the difference was almost an order of magnitude (Fig. 2). Both membranes revealed a decrease of amplitude in the frequency range

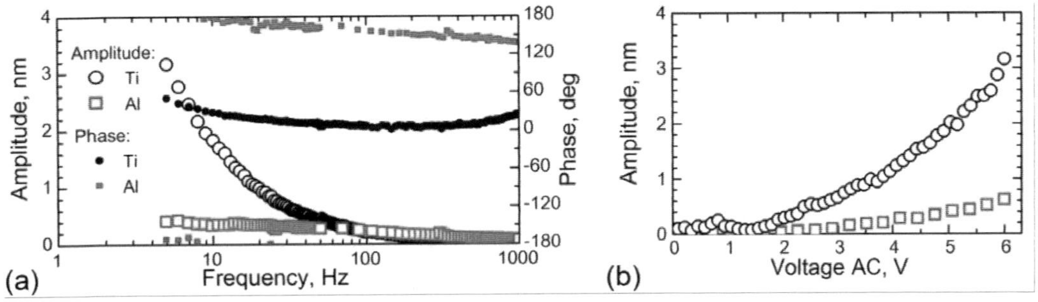

Figure 2. (a) Frequency (under 6 V AC) and (b) AC voltage (at 5 kHz) dependences of 2nd harmonic EM response in membranes with Al and Ti electrodes.

1Hz–1kHz. The phase of the EM response measured by the lock-in technique was different for different membranes. Membrane with Ti electrodes revealed close to zero value, while the one with Al electrode had a phase lag about 120-180 degrees with respect to the excitation voltage. The phase lag was also decreasing with frequency for both types of the membranes.

The increase of EM amplitude at low frequencies allowed measuring displacement-voltage dependences under slowly changed saw-type voltage (Fig. 3). The membranes with Al electrodes showed weak frequency dependence in the range of 2-50 Hz and almost temperature-independent behavior. The response was strongly asymmetric being about 30% lower, if the negative potential was applied to the top electrode. The asymmetry also depended on frequency. On the contrary, membranes with Ti electrodes demonstrated strong decrease of the EM signal with frequency (about an order in the studied frequency range). Moreover, membranes with Ti electrodes demonstrated strain hysteresis, while in the membranes with Al contacts hysteresis was observed only at high applied driving voltage.

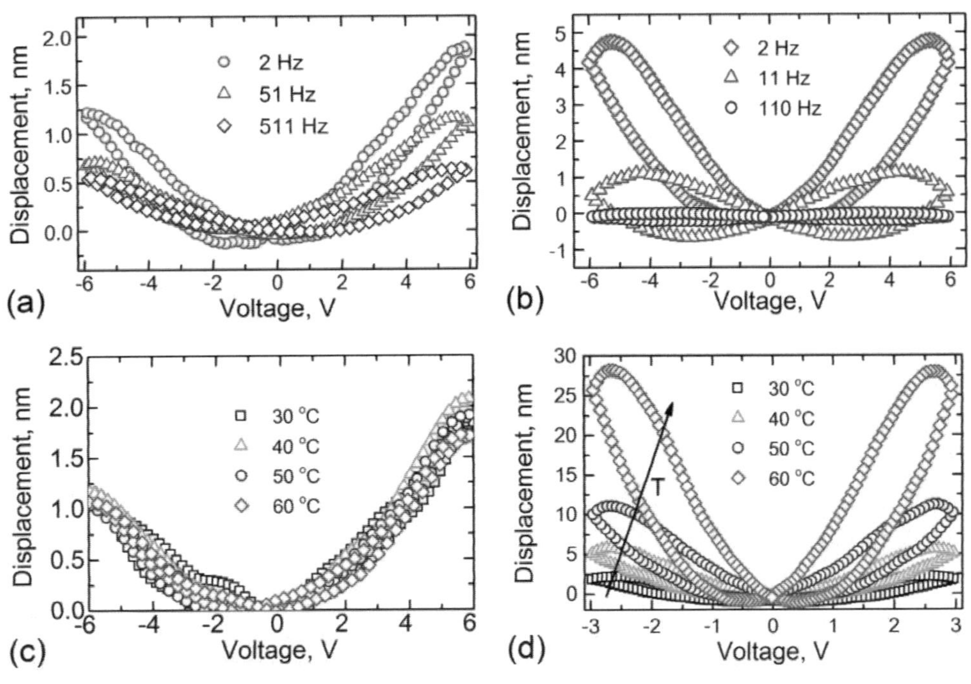

Figure 3. (a,b) Frequency and (c,d) temperature dependent displacement-voltage loops of the CGO membranes with (a, c) Al and (b, d) Ti electrodes.

IOP Conf. Series: Materials Science and Engineering **256** (2017) 012008 doi:10.1088/1757-899X/256/1/012008

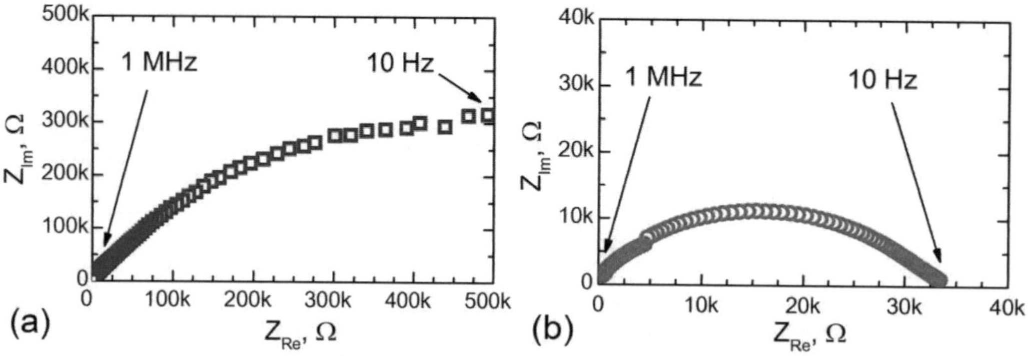

Figure 4. Cole-Cole diagrams of membranes with (a) Al and (b) Ti electrodes.

We used the impedance spectroscopy to verify the contact quality and electric field distribution (Fig. 4) within the membranes. The impedance spectra of membrane with Al electrode demonstrated clear Schottky-barrier behavior at low frequencies [13]. The existence of the Schottky-barrier resulted in voltage drop at the CGO/Al interface and additional barrier resistance. The impedance of the films was about 120 Ohms at 1 MHz under 5 V applied voltage, which implies that the dissipated power was about 200 mW. The dissipated power can be a source of heating by a few degrees, which can result in detectable strain due to thermal expansion. Both electrostriction and Joule heating are proportional to the electric field squared, which makes difficult to distinguish them.

The appearance of the Schottky barrier in CGO with Al contacts can explain the difference in displacement-voltage dependences (Fig. 3). Indeed, the blocking nature of the Al contacts dominates the overall resistance at all frequencies and temperatures.

Therefore, we decided to use amplitude modulated signal with 1MHz carrier to probe the response of the membranes at different frequencies. In the case of amplitude modulation, the electric field is equal to:

$$E_{AM} = A \cdot \cos(\omega_c t)\,(1 + M \cdot \cos(\omega_m t)), \tag{1}$$

where A is the amplitude of carrier signal, M is the amplitude of modulation, ω_c is the carrier frequency, and ω_m is a modulation frequency. For modulation depth of 100% (A equals to M) the strain may now be expressed as:

$$S = \frac{dL}{L} = ME^2 = M \cdot \left(A \cdot \cos(\omega_c t) + \frac{AM}{2}(\cos(\omega_c t - \omega_m t) + \cos(\omega_c t + \omega_m t))\right)^2 = \\ = \frac{3MA^2}{4} + MA^2\cos(\omega_m t) + \frac{MA^2}{4}\cos(2\omega_m t) + \frac{3MA^2}{4}\cos(2\omega_c t) + \cdots, \tag{2}$$

where dL is the absolute deformation of sample, L is the thickness of sample, M is the electrostriction coefficient. Thus, the electrostriction response could be observed on the first harmonic of the excitation.

The measurements of the EM response at the 1st harmonics were done in the 100% modulation regime (modulation coefficient $h = 1$) in a modulation frequency range 5Hz-1kHz and voltage amplitude up to 1 V (Fig. 5, black line). Amplitudes of normalized displacements from both membranes were higher than those in low frequency; however, in the case of Al electrodes this increase was more significant (Fig. 5). Analysis of the phase allowed us to make an assumption about important contribution into EM response at high and low frequencies. Phase lag in the case of electrostriction effect is supposed to be close to zero, because it is, generally, the elastic response, while Joule heating must have retardation due to a limited cooling rate. In the case of Al electrode and low frequency, we expect no electrostriction due to the Schottky barrier at the electrode. Indeed, we have in that case the highest value of the phase lag manifesting the general Joule heating input in the

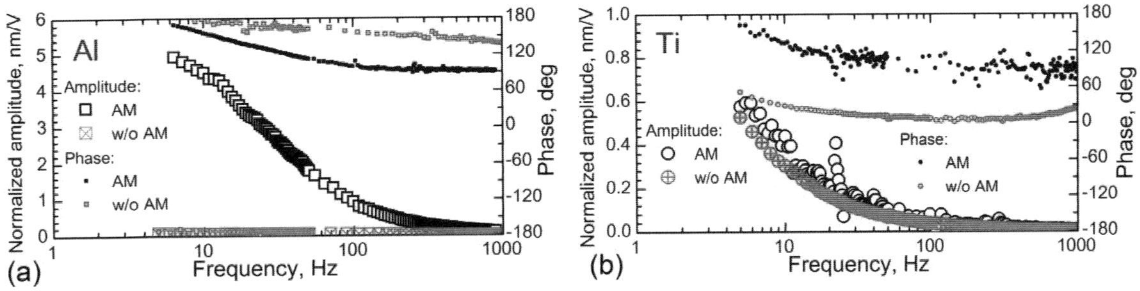

Figure 5. Frequency dependences of EM response depend on amplitude modulation (black) or 2nd harmonic detection (red, 6 V driving voltage) regimes of membranes with (a) Al and (b) Ti electrodes. 1 MHz carrier frequency and 1 V of voltage amplitude were used for amplitude modulation regime.

EM response. On the contrary, the membrane with Ti electrode demonstrated the minimal phase lag at low frequency, likely, because of the domination of the electrostriction. At high frequency, the phase lag became close to 60 degrees for both membranes indicating mixing of the signals from electrostriction and Joule heating. Actually, Al electrodes became non-blocking for the high frequency range of applied voltages, while membrane with Ti electrode at high frequency became heated due to increase of its resistance. That is why the mixed regime was realized in a high frequency range and phase lag became similar for both membranes.

Thus, the electrode providing the voltage for the membrane significantly limited the frequency range, where the CGO self-supported membrane could be operated in a regime with dominant electrostriction effect without essential heating.

4. Conclusion

We demonstrated the existence of two principal excitation mechanisms of the self-supported Gd-doped cerium oxide $Ce_{0.8}Gd_{0.2}O_{1.95}$ (CGO) thin circular membranes: electrostriction effect and Joule heating. Al electrodes apparently formed a blocking interface with CGO, probably a Schottky barrier. Therefore, at low frequencies no field was applied at CGO and at high frequencies, the barrier became electrically transparent because of the high electrical capacitance. As a result, at low frequencies the response of the membranes with Al contacts was very small, but at high frequencies it was dominated by the Joule heating. The resistance of the titanium electrodes was much smaller and a large fraction of the electric field fell on the CGO layer already at low frequencies. In this case, the electrostriction became a dominant excitation mechanism. Operating the membranes at high frequencies (> 0.1 MHz) caused a mixed regime of excitation.

Acknowledgements

The equipment of the Ural Center for Shared Use "Modern Nanotechnology" UrFU has been used. The research was made possible in part by RFBR (Grant No. 15-52-06006 MNTI_a). This work was supported by the Israeli Ministry of Science and Technology within the program of Israel-Russian Federation Scientific Collaboration, Grant No. 12421-3. This work has been supported in part by the Ministry of Education and Science of the Russian Federation under Project No. 3.9534.2017/BP.

References

[1] Sun E and Cao W 2014 *Prog. Mater. Sci.* **65** 124-210
[2] Zhang S and Li F 2012 *J. Appl. Phys.* **111** 031301
[3] Grant L D 2009 Lead and compounds *Environmental Toxicants: Human Exposures and Their Health Effects (3rd ed.)* ed M Lippmann (New York: Wiley-Interscience) ch. 20 pp 757-809
[4] Järup L 2003 *Brit. Med. Bull.* **68** 167-82
[5] Yavo N, Smith A D, Yeheskel O, Cohen S, Korobko R, Wachtel E, Slater P and Lubomirsky I

2016 *Adv. Funct. Mater.* **26** 1138-42

[6] Kalinin S V and Gruverman A 2007 *Scanning Probe Microscopy Electrical and Electromechanical Phenomena at the Nanoscale* (New York: Springer-Verlag)

[7] Korobko R, Patlolla A, Kossoy A, Wachtel E, Tuller H L, Frenkel A I and Lubomirsky I 2012 *Adv. Mater.* **24** 5857-61

[8] Newnham R E, Sundar V, Yimnirun R, Su J and Zhang Q M 1997 *J. Phys. Chem.* B **101** 10141-50

[9] Mishuk E, Makagon E, Wachtel E, Cohen S R, Popovitz-Biro R and Lubomirsky I 2017 *Sensor. Actuat. A-Phys.* **264** 333-40

[10] Ushakov A D, Mishuk E, Makagon E, Alikin D O, Esin A A, Baturin I S, Tselev A, Shur V Ya, Lubomirsky I and Kholkin A L 2017 *Appl. Phys. Lett.* **110** 142902

[11] Ushakov A D, Yavo N, Mishuk E, Lubomirsky I, Shur V Ya and Kholkin A L 2016 *Proc. IV Sino-Russian ASRTU Symp. on Advanced Materials and Materials and Processing Technology (Ekaterinburg)* (Dubai: KnE Materials Science) p 177

[12] Yimnirun R, Moses P J, Meyer Jr R J and Newnham R E 2003 *Meas. Sci. Technol.* **14** 766-72

[13] Lunkenheimer P 2002 *Phys. Rev.* B **66** 052105

The study of electrical conductivity of DNA molecules by scanning tunneling spectroscopy

T I Sharipov* and R Z Bakhtizin

Department of Physical Electronics and Nanophysics, Bashkir State University, 450076 Ufa, Russia

*sha-t@yandex.ru

Abstract. An interest to the processes of charge transport in DNA molecules is very high, due to perspective of their using in nanoelectronics. The original sample preparation for studying electrical conductivity of DNA molecules by scanning tunneling spectroscopy has been proposed and tested. The DNA molecules immobilized on gold surface have been imaged clearly and their current-voltage curves have been measured.

1. Introduction

The use of DNA molecules in nanoelectronic devices is very promising [1, 2] since it may cause significant interest to the processes of charge transport in these molecules. Although the double helix DNA form was discovered more than 60 years ago [3], the correct relationship between its structural, chemical, and electrical properties is not established yet. One of the problems is that attempts to measure the electrical resistance of DNA resulted in contradictory conclusions: the molecule can demonstrate dielectric [4, 5], semiconductor [6], conductor [7-9], and even superconductor [10] properties. Fink and Schonenberger [7] have shown that current-voltage curves of λ–DNA molecules of 1 μm length are linear and their resistivity $\rho \sim 10^{-4}\ \Omega\cdot$cm. Porath et al. [6] have described current-voltage curves with a clear gap of around 2 V and a resistance of 3 GΩ at a voltage of 4 V for Poly(G)-Poly(C) DNA sequence length of 10 nm ($\rho \sim 10\ \Omega\cdot$cm). The reasons of such results ambiguity can be attributed to a complex structure of the DNA molecule, as well as uncertainty in treatment of the experimental results. In addition, the sample preparation and research methodology that influence the conductivity mechanism in DNA are also important.

There are two main approaches for conductivity research of the DNA molecule. First, the isolated DNA molecule is placed between two electric contacts [11-13]. The distance between electrodes should be small (about 1-10 nm), but sufficient enough to interfere with tunneling of charges. Such experiment is rather hard to carry out and extremely difficult to control.

Second approach is based on the use of a scanning tunneling microscope (STM) [14, 15]. First experiments with STM have shown that it is the most suitable research tool for both single DNA molecules and molecules in monolayer films [16, 17]. However, qualitative experimental measurements of charge transport through a single molecule by means of STM as well as interpretation of the results are extremely difficult. A great experimental problem is arranging the contacts to a single molecule. The accuracy of such manipulation is very important for charge transport.

IOP Conf. Series: Materials Science and Engineering **256** (2017) 012009 doi:10.1088/1757-899X/256/1/012009

2. Materials and methods

We have used scanning probe microscope Solver P47 and Probe nanolaboratory Ntegra-Prima («NT-MDT», Zelenograd, Russia) for the AFM/STM study and for scanning tunneling spectroscopy (STS) of DNA molecules. All investigations were carried out in air. The deionized distilled water MilliQ (Millipore, France) was used in preparation of all solutions.

Specimens were 20-mer oligonucleotide ("Syntol", Moscow, Russia) and the 5-thiol modified oligonucleotides. The following reagents were used also: 1 mM 2-mercaptoethanol, the hybridization buffer is 100 mM Tris.HCl/100mM NaCl, the solution for washing 100 mM Tris HCl/300 mM NaCl.

A plate of silicon Si (111) alloyed with phosphorus with a specific resistance of 1.6 Ω·cm (the nominal concentration of the current carriers ~ $8\cdot10^{17}$ cm^{-3}) was used as an initial substrate material. After degreasing in toluene, the silicon substrates have been washed in ultrasonic bath with acetone, and then by deionized water. Next, we have carried out the thermal deposition of gold onto the silicon substrate.

The next stage of the experiment was the modification of the gold surface to bind with the studied oligonucleotides. Chemical modification included the creation of the covalent bond of chemically polarized groups of thiols with a clean metal surface. Adsorption of DNA was carried out by the Coulomb interaction of a molecule with a tightly packed monomolecular film of thiol oriented by its positive functional groups to the DNA molecule.

5'-thiol modified oligonucleotides were dissolved in water. Then, 5 μl of this solution was dropped on a gold substrate, which was placed in a Petri dish with moist atmosphere, created using the cuvette with water and aged for 16 hours at 50°C. After that, the substrate was washed twice by water and a solution of 2-mercaptoethanol. This solution allowed increasing the availability of immobilizing samples to complementary sequences. Finally, the substrate was placed back into the Petri dish, kept for 2 hours at 40·°C, and washed by water.

Oligonucleotides in the concentration of 1 μM were placed for 10 minutes in the solution for hybridization at 80·°C. Immediately after this process, 5 μl of this solution was dropped on the substrate, which was then incubated for 1 hour in air at 40·°C. Then it was washed in the buffer for rinsing and afterwards with clean water. The prepared substrates were dried at room conditions.

3. Results and discussion

AFM study revealed DNA molecules rarely located on the sample surface (Fig. 1a) [18]. We needed to obtain the STM image for identification of DNA molecules on the substrate in order to measure the current-voltage curves. It is known that the DNA molecules are represented as the dark spots in the STM image, due to lower electrical conductivity as compared to gold.

STM-study of the gold surface with immobilized DNA molecules was performed at the constant tunneling current mode. On the STM image (Fig. 1b), dark objects of the small diameter – DNA molecules are observed with lateral dimensions similar to the dimensions of DNA on the AFM image.

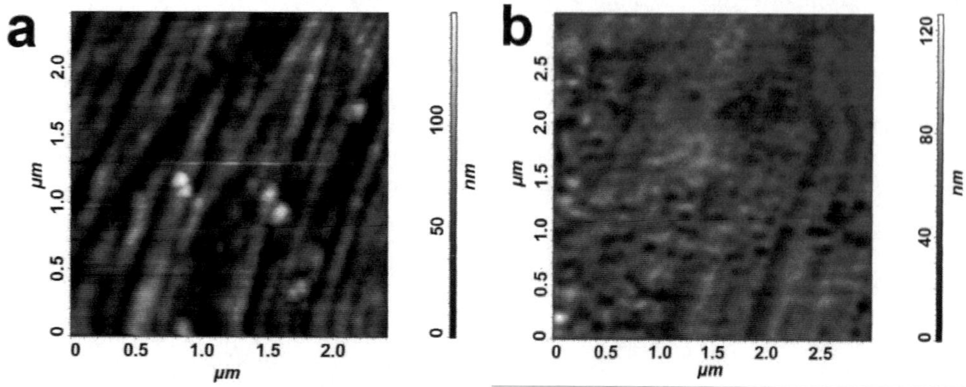

Figure 1. Images of DNA molecules on the gold surface: (a) AFM, (b) STM.

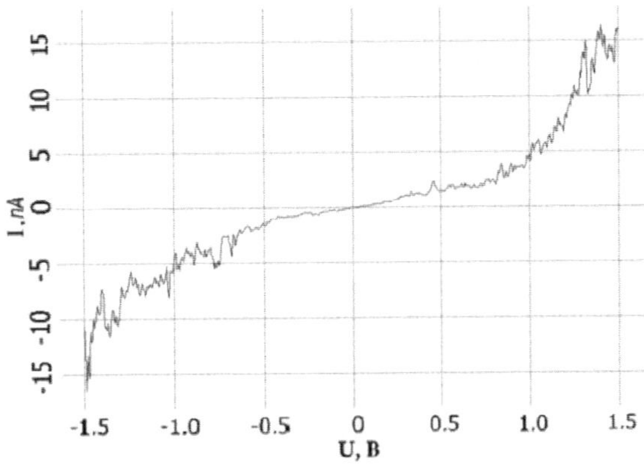

Figure 2. Current-voltage curve of DNA molecules.

The DNA molecules were identified on STM images and the current-voltage curves were measured by scanning tunneling spectroscopy mode in dark locations (Fig. 2). The voltage U applied between the gold substrate and the STM tip ranged from -1.5 V to +1.5 V.

The symmetric current-voltage curve with pronounced manifestation of nonlinearity and the growth of a tunnel current dispersion of fluctuations have been observed. The obtained current-voltage curves were compared with the approximated ones for semiconductor, dielectric, and conductor. We have found out that our approximation curves have the most similarity with the current-voltage curves of the semiconductor.

4. Conclusion

For investigating the dependence of the tunneling current through the DNA molecule on the magnitude of the applied voltage between the gold substrate and the STM tip, the current-voltage curves have been measured by scanning tunneling spectroscopy method. The study of electrical conductivity has crucial importance in the development of biosensors.

Acknowledgments

This work was supported by Russian Foundation for Basic Research under grant No. 17-42-020616.

References

[1] Zalis S, Kratochvílová I, Zambova A, Mbindyo J, Mallouk T and Mayer T 2005 *Eur.Phys.J.* E **18** 201–6
[2] Kratochvílová I, Nespurek S, Sebera J, Zális S, Pavelka M, Wang G and Sworakowski J 2008 *Eur.Phys.J.* E **25** 299–307
[3] Watson J D and Crick F H 1953 *Nature* **171** 737-8
[4] Brauns E B, Madaras M L, Coleman R S, Murphy C J and Berg M A 1999 *J.Am.Chem.Soc.* **121** 11644
[5] Storm A J, van Noort J, de Vries S and Dekker C 2001 *Appl. Phys. Lett.* **79** 3881
[6] Porath D, Bezryadin A, de Vries S and Dekker C 2000 *Nature London.* **403** 635
[7] Fink H W and Schonenberger C 1999 *Nature* **398** 407
[8] Rakitin A, Aich P, Papadopoulos C, Kobzar Y, Vedeneev A S, Lee J S and Xu J M 2001 *Phys. Rev. Lett.* **86** 3670
[9] Tran P, Alavi B and Grüner G 2000 *Phys. Rev. Lett.* **85** 1564
[10] Kasumov A Y, Kociak M, Gueron S, Reulet B, Volkov V T, Klinov D V and Bouchiat H 2001 *Science* **291** 280
[11] Morpurgo A F, Marcus C M and Robinson D B 1999 *Appl. Phys. Lett.* **74** 2084
[12] de Pablo P J, Moreno-Herrero F and Artacho E 2001 *Appl. Phys. Lett.* **79** 3881

[13] de Pablo P J, Moreno-Herrero F, Colchero J, Gomez Herrero J, Herrero P, Baro A M, Ordejon P, Soler J M and Artacho E 2000 *Phys. Rev. Lett.* **85** 4992-5
[14] Cohen H, Nogues C, Naaman R Porath D 2005 *PNAS* **102** 11589-93
[15] Kratochvílova I, Kral K, Buncek M et al. 2008 *Biophys. Chemistry* **138** 3-10
[16] Onipko A I, Berggren K F, Klymenko Y O, Malysheva L I, Rosnik J J W M, Geerlings L J, van der Drift E and Radelaar S 2000 *Phys. Rev.* B **61** 11118–24
[17] Rosnik J J W M, Blauw M A, Geerlings L J, van der Drift E and Radelaar S 2000 *Phys. Rev.* B **62** 10459–66
[18] Sharipov T I and Bakhtizin R Z 2017 *IOP Conf. Series: Materials Science and Engineering* **195** 012002 doi:10.1088/1757-899X/195/1/012002

IOP Conf. Series: Materials Science and Engineering **256** (2017) 012010 doi:10.1088/1757-899X/256/1/012010

Peculiarities of living cell response to the external stimuli revealed via quasistatic mode of atomic force microscopy

M M Khalisov[1,2,*], A V Ankudinov[2,3], V A Penniyaynen[1], T E Timoshenko[1], K I Timoshchuk[2], M V Samsonov[4] and V P Shirinsky[4]

[1] Pavlov Institute of Physiology RAS, Saint Petersburg, 199034, Russia
[2] ITMO University, Saint Petersburg, 197101, Russia
[3] Ioffe Institute, Saint Petersburg, 194021, Russia
[4] National Cardiology Medical Research Center, Moscow, 121552, Russia

*hamax@list.ru

Abstract. The technique of atomic force microscopy allows revealing living cell morphology and mechanical properties characterization under physiologically relevant conditions. Here, we review our recent results on living cell reaction to different external influences obtained by this technique. The Bruker PeakForce QNM quasistatic mode was used to study living fibroblasts, erythrocytes, sensory neurons, and endothelial cells.

1. Introduction

In the last decades, atomic force microscopy (AFM) has become one of the common tools in cell biology research. Now AFM is widely used for living animal cell characterization [1, 2]. This imaging method allows studying cell morphology and mechanical properties with submicrometer resolution and under physiologically relevant conditions.

Various cellular processes are associated with mechanical properties of cells: shape maintenance, migration, differentiation, and division [3]. Several pathologies are known to change the cell mechanical phenotype [4]. Although the living cells are not perfectly elastic objects, in AFM studies their mechanical properties are mostly described by Young's modulus. The measurement of the Young's modulus of a single cell is based on the AFM nanoindentation technique. It implies that the AFM probe compresses the sample with a preset force, and the induced cell deformation is employed to calculate the Young's modulus according to the selected contact mechanics model [3].

AFM Force Spectroscopy and Force Volume are the most widespread modes for living cell Young's modulus measurements. These modes are slow in operation and have low data flow rate. Recently developed quasistatic AFM modes, such as PeakForce QNM (Bruker), are much faster; therefore, they appear to be promising for living cell examination. PeakForce QNM and other similar modes force the AFM tip oscillate well below its resonance frequency and press against the sample surface for a short period of time. The manner of operation enables simultaneous topography and mechanical properties data acquisition. Since PeakForce QNM mode was integrated into commercial AFM systems not long ago, the information on its application to living cells is pretty scarce [5-7]. We utilized AFM in PeakForce QNM mode to study the mechanical properties of different types of living cells. The aim of this work was to examine and analyze the specifics of AFM investigation of living chicken embryo cardiac fibroblasts, rat erythrocytes, chicken embryo sensory neurons, and mouse microvascular endothelial cells under physiologically relevant conditions.

IOP Conf. Series: Materials Science and Engineering **256** (2017) 012010 doi:10.1088/1757-899X/256/1/012010

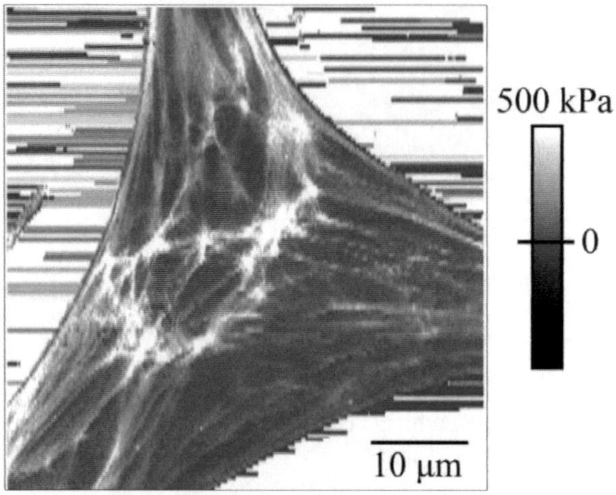

Figure 1. AFM map of Young's modulus of living cardiac fibroblast obtained using a sharp tip. The AFM imaging parameters: peak force – 3 nN, probing amplitude and vertical frequency – 1.0 μm and 250 Hz, horizontal scanning frequency – 0.1 Hz.

2. Materials and methods

All experiments were performed using the atomic force microscope (Bruker BioScope Catalyst) combined with the inverted optical microscope (Carl Zeiss Axio Observer). During the AFM imaging in PeakForce QNM mode, the cells were in physiological solution at 37°C. AFM data were processed in NanoScope Analysis 1.50 and Gwyddion 2.44.

We studied cardiac fibroblasts of 10-12-day-old chicken embryos using two types of AFM probes: conventional sharp tips ($R \leq 12$ nm) and colloidal indenters ($R = 325$ nm) [8]. The information on the sample handling is found in [9].

3. Results and discussion

As expected, sharp probes enabled more detailed AFM images. As a consequence, Young's modulus maps permitted single cytoskeletal filaments identification in the submembrane layer of cell (Fig. 1).

The apparent Young's moduli of fibroblast determined using Sneddon (sharp tips) and Hertz (colloidal probes) models differed considerably: $E_{Sneddon}/E_{Hertz} \approx 5$. At the same time, the mean of cell maximum height and the contact stiffness were similar when using sharp as well as submicron spherical indenters: ≈ 1.7 μm and ≈ 18 mN/m, respectively. The result of our observations implies that, in PeakForce QNM AFM, Young's modulus is inaccurate to quantify fibroblast mechanical properties; contact stiffness is preferable instead. Since the fibroblast contact stiffness did not depend on the AFM tip shape, we suppose that the external cell layer behaves as a hard shell and its deformation is determined only by the applied load.

The next object studied is a living erythrocyte or red blood cell. Extreme deformability of living erythrocytes prevents their effective AFM testing. Therefore, these cells are often examined in the hardened fixed state. In practice, such a state is attained by erythrocyte chemical treatment or drying, but fixing precludes analysis of cell native elastic properties, which are of great interest. In order to overcome this limitation, living erythrocytes are immobilized on adhesive substrate polylysine. A few examples of AFM investigations of living red blood cells on polylysine may be found in [10–13]. Yet polylysine was reported to be able to modify erythrocyte properties and even to invoke lysis. The authors [14] stated that erythrocyte deposition on polylysine led to the shift of cell shape to a spherical cap following by the plasma membrane damage due to the sustained static membrane tension. Optical observation indeed confirmed these conclusions: over time living erythrocytes "vanished".

We imaged living rat erythrocytes attached to polylysine in PeakForce QNM mode of AFM. Sample preparation procedure is presented in [15]. Using optical microscope, we also detected disappearance of some cells with the passage of time. However, in contrast to Hategan et al. [14], our findings show that erythrocytes do not actually vanish; instead, they change their optical contrast and shape as well as rigidity. Firstly, after the deposition red blood cells spread on the polylysine substrate.

In this case, the cells are flat and highly deformable. Besides, they possess optical contrast relative to the adhesive layer. Subsequently, erythrocytes swell, becoming convex and stiff. Moreover, the cells simultaneously lose their color when monitoring in the optical microscope. That is, as decolorized erythrocytes are visualized via AFM, we conclude that they maintain their integrity during the discovered transformation. In one AFM experiment, we have managed to visualize the erythrocyte transformation process (Fig. 2).

We have shown that living erythrocytes tend to transform on the polylysine substrate. The erythrocyte state characterized by cell swelling and hardening precedes lysis. In general, our findings indicate that the plasma membrane extra tension induced by cell-substrate strong adhesion should be taken in consideration while performing living cell AFM studies.

AFM was also employed to study the living sensory neuron response to substances with analgesic effect. Comenic acid (CA) was shown to relieve chronic pain [16]. This chemical has become an active ingredient of the novel non-opioid painkiller "Anoceptin" [17]. Another promising substance for analgesia is ouabain in nanomolar concentration, which is capable to decrease the voltage sensitivity of sensory neuron slow sodium $Na_v1.8$ channels responsible for pain signals in mammals [18, 19]. We used AFM to clarify, if CA and ouabain affect the morphology and elastic properties of living sensory neurons.

The AFM sample preparation protocol one can find in [20]. Dorsal root ganglia neurons of 10-12-day chicken embryos were cultivated on polylysine in the presence of CA (10^{-8} M) or ouabain (10^{-10} M). Using AFM we investigated the morphology and Young's modulus of living neurons of three groups: control cells, CA- and ouabain-treated neurons.

No morphology changes were detected in living sensory neurons under the influence of CA and ouabain. The control and CA-treated neurons manifested close mean Young's modulus values: 115 ± 90 kPa ($n = 30$) and 120 ± 97 kPa ($n = 21$), respectively. However, the mean neuron Young's modulus of cells after ouabain treatment was somewhat higher: 172 ± 109 kPa ($n = 15$). Although the ouabain-treated sensory neuron elasticity surpassed the control cell elasticity by a factor of approximately 1.5, Mann-Whitney U-test showed no statistically significant difference between two Young's modulus values ($p = 0.08$).

Interestingly, earlier we observed the similar effect of ouabain on sensory neurons cultivated on collagen-fibronectin-coated Petri dishes [21]. The same trend of neuron stiffening after ouabain treatment on two diverse substrate types verifies the consistency of our results. The fact, that ouabain stiffens sensory neurons and CA does not, testifies to the substances activate signaling pathways that differ from each other.

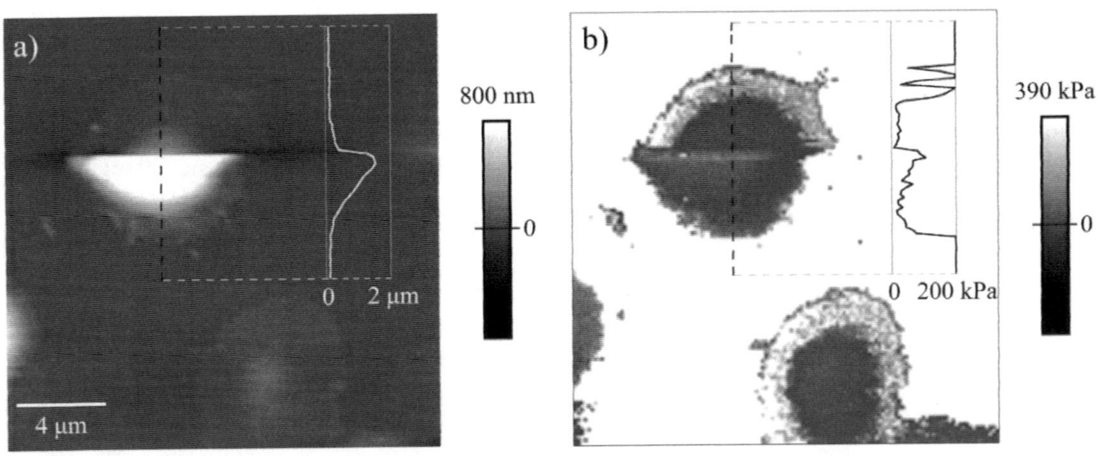

Figure 2. AFM (a) topography and (b) Young's modulus map of living rat erythrocytes. AFM imaging parameters: peak force 0.5 nN; probing amplitude and vertical frequency 1.0 µm and 250 Hz; horizontal scanning frequency 0.1 Hz; slow scan direction – from top to bottom.

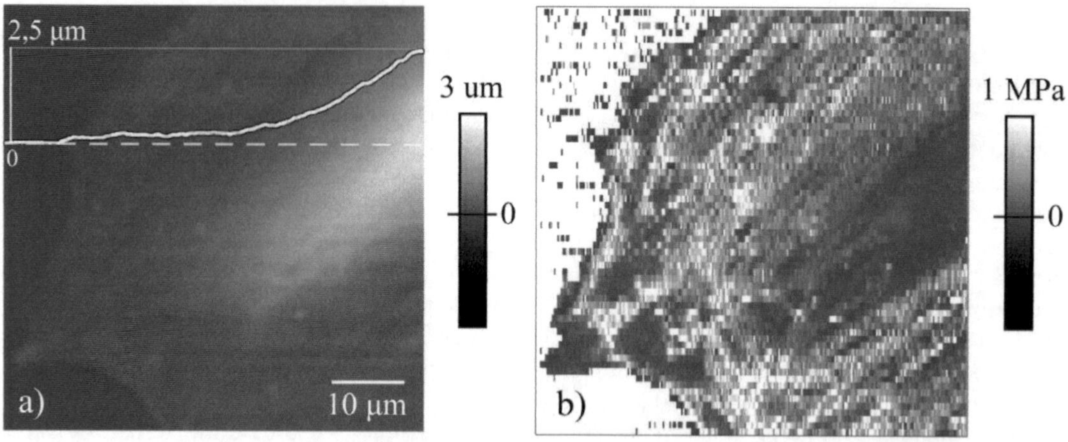

Figure 3. AFM EC (a) height image and (b) Young's modulus map. AFM imaging parameters: peak force 1 nN; probing amplitude and vertical frequency 0.5 μm and 250 Hz; horizontal scanning frequency 0.15 Hz.

AFM PeakForce QNM millisecond indentation thus allows tracing changes in Young's modulus of sensory neuron values in response to addition of chemical agents with analgesic effect on collagen-fibronectin or polylysine substrate. In the case of polylysine, ouabain induces a 1.5-fold increase of mean Young's modulus of sensory neuron.

The next part of our investigation was dedicated to the cells of endothelium. The monolayer of the endothelial cells (ECs) forms continuous inner surface of blood vessels. The EC mechanical properties are of interest, because they are considered important for blood pressure regulation and cardiovascular disease progression. The contribution of different macromolecules and signaling events to EC mechanical phenotype is still poorly understood. Previous studies suggested the active role of actin cytoskeleton and regulatory proteins such as myosin light chain kinase in control of cellular stiffness and mechanosensing [22, 23]. According to the accomplished measurements, height distribution across EC is heterogeneous: cell center has the maximum thickness (~ 1 μm) due to the presence of nucleus, while the distal regions are significantly thinner (~ 100 nm) (Fig. 3a). Young's modulus of EC (Fig. 3b) apparently mainly reflects the nucleus mechanical properties. Probing the EC elasticity at cell periphery allows avoiding the nucleus effect. However, the cell periphery is too thin and its Young's modulus quantification could potentially be hampered by the proximity of the stiff substrate. Taking into account the typical cell indentation depth in PeakForce QNM mode (~ 100 nm), it was important to clarify whether the hard substrate would conceal the difference in mechanical properties of EC thin regions under various EC treatment conditions. In order to test this assumption, we compared the EC distal region Young's moduli of normal, latrunculin B-treated, and nocodazole-treated cells. These substances inhibit the formation of the essential components of cytoskeleton, microfilaments, and microtubules, respectively.

Our experiments showed that living mouse lung microvascular ECs grown on the gelatin substrate have large lateral dimensions and tend to form clusters. As individual ECs or cell clusters usually surpass AFM frame size, the reliable local cell height determination was problematic. Hence, we could not apply the Gavara-Chadwick correction method to derive precise Young's modulus values [24].

Young's modulus of EC periphery was analyzed within an area between cell nucleus and cell edge. It turned out that the experimental Young's modulus histogram shape best corresponded to the log-normal distribution. For this reason, we employed median Young's modulus values to characterize ECs.

The comparison of averaged median Young's modulus values of EC distal regions has led to the following results: control untreated cells exhibited the highest Young's modulus value, 140 ± 100 kPa (n = 26), whereas all inhibitors induced softening of ECs. The cells treated by latrunculin B and nocodazole had the Young's modulus values of 60 ± 60 kPa (n = 8) and 70 ± 60 kPa (n = 25), respectively.

Statistical significance of our results was estimated using Mann–Whitney U test [25, 26]. P-values less than 0.05 were considered statistically significant. P-values were as follows: $p = 0.017$ for latrunculin B and $p = 0.003$ for nocodazole. Thus, ECs after latrunculin B and nocodazole treatment showed statistically significant change of Young's modulus.

The disassembly of microfilaments and microtubules resulted in reduction of EC periphery Young's modulus. According to our experimental data, the most drastic decrease of Young's modulus was observed in latrunculin B-treated ECs. It supports the view that actin component of cytoskeleton is the main contributor to cell elasticity [27].

Thus, our findings allow concluding that AFM indentation of thin distal regions of EC in PeakForce QNM protocol does not conceal the detection of cell stiffness alterations induced by specific inhibitors of actin and tubulin polymerization. Therefore, one can use PeakForce QNM indentation to reveal how different protein molecules affect mechanical properties of EC cortical cytoplasm.

4. Conclusion

A quasistatic AFM mode such as PeakForce QNM allows measuring the mechanical properties of living cells. Various cell types display peculiarities in mechanophenotype when subjected to external stimuli. This should be taken into account when designing high throughput AFM screening protocols of living cells for biomedical applications.

Acknowledgments

The authors are indebted to professor B.V. Krylov for his interest to our work. M.M. Khalisov and A.V. Ankudinov are grateful for the support from the Government of the Russian Federation (Grant No. 074-U01).

References

[1] Kuznetsova T G, Starodubtseva M N, Yegorenkov N I, Chizhik S A and Zhdanov R I 2007 *Micron.* **38** 824–33

[2] Efremov Yu M, Bagrov D V, Dubrovin E V, Shaitan K V and Yaminskii I V 2011 *Biofizika* **56** 288–303

[3] Gavara N 2017 *Microsc Res Tech.* **80** 75-84

[4] Rianna C and Radmacher M 2016 *AIP Conference Proceedings.* **1760** 020057-1–13

[5] Tomankova K, Kolar P, Malohlava J and Kolarova H 2012 *Current Microscopy Contributions to Advances in Science and Technology (A. Méndez-Vilas, Ed.).* **1** 549–54

[6] Rigato A, Rico F, Eghiaian F, Piel M and Scheuring S 2015 *ACS Nano.* **9** 5846–56

[7] Calzado-Martín A, Encinar M, Tamayo J, Calleja M and San Paulo A 2016 *ACS Nano.* **10** 3365–74

[8] Nyapshaev I A, Ankudinov A V, Stovpyaga A V, Trofimova E Y and Eropkin M Y 2012 *Tech Phys.* **57** 1430–7

[9] Khalisov M M, Ankudinov A V, Penniyaynen V A, Nyapshaev I A, Kipenko A V, Timoshchuk K I, Podzorova S A and Krylov B V 2017 *Tech Phys Lett.* **43** 209–12

[10] Mozhanova A A, Nurgazizov N I and Bukharaev A A 2003 *SPM-2003. Proc. Nizhni Novgorod.* 266–7

[11] Lekka M, Fornal M, Pyka-Fosciak G, Lebed K, Wizner B, Grodzicki T and Styczen J 2005 *Biorheology.* **42** 307–17

[12] Maciaszek J L, Andemariam B and Lykotrafitis G J 2011 *J Strain Anal Eng.* **46** 368–79

[13] Rebelo L M, Sousa J S, Santiago T M and Mendes Filho J 2014 *Microscopy: advances in scientific research and education.* **1** 141–52

[14] Hategan A, Law R, Kahn S and Discher D E 2003 *Biophys J.* **85** 2746–59

IOP Conf. Series: Materials Science and Engineering **256** (2017) 012010 doi:10.1088/1757-899X/256/1/012010

[15] Khalisov M M, Timoshchuk K I, Ankudinov A V and Timoshenko T E 2017 *Tech Phys.* **62** 310–3
[16] Plakhova V, Rogachevsky I, Lopatina E, Shelykh T, Butkevich I, Mikhailenko V, Otellin V, Podzorova S and Krylov B 2014 *Act Nerv Super Rediviva.* **56** 55–64
[17] Krylov B, Rogachevsky I and Plakhova V Substance with sedative effect 2006 US 20060252824 A1
[18] Borovikova L V, Borovikov D V, Ermishkin V V and Revenko S V 1997 *Primary Sensory Neuron.* **2** 65–75
[19] Gold M S, Reichling D B, Shuster M J and Levine J D *Proc Natl AcadSci USA.* 1996 **93** 1108−12
[20] Khalisov M M, Penniyaynen V A, Esikova N A, Ankudinov A V and Krylov B V 2017 *Tech Phys Lett.* **43** 85–7
[21] Ankudinov A V, Khalisov M M, Penniyainen V A, Podzorova S A and Krylov B V 2015 *Tech Phys.* **60** 1540–4
[22] Cai S, Pestic-Dragovich L, O'Donnell M E, Wang N, Ingber D, Elson E and De Lanerolle P 1998 *Am J Physiol.* **275** 1349–56
[23] Ohlmann P, Tesse A, Loichot C, RalayRanaivo H, Roul G, Philippe C, Watterson D M, Haiech J and Andriantsitohaina R 2005 *Am J Physiol Heart Circ Physiol.* **289** 2342–9
[24] Gavara N and Chadwick RS 2012 *Nat Nanotechnol.* **7** 733–6
[25] Vargas-Pinto R, Gong H, Vahabikashi A and Johnson M 2013 *Biophys J.* **105** 300–9
[26] Jalilian I, Heu C, Cheng H, Freittag H, Desouza M, Stehn J R, Bryce N S, Whan R M, Hardeman E C, Fath T, Schevzov G and Gunning P W 2015 *PLoS One.* **10** 1–23
[27] Rotsch C and Radmacher M 2000 *Biophys J.* **78** 520–35

IOP Conf. Series: Materials Science and Engineering **256** (2017) 012011 doi:10.1088/1757-899X/256/1/012011

Formation features and AFM studies of the silicon-chitosan-containing glycerohydrogel surface

Yu Yu Zhuravleva*, O N Malinkina, D N Bratashov and A B Shipovskaya

Chernyshevsky Saratov State University, 410012 Saratov, Russia

*zhuravleva11.03@mail.ru

Abstract. Hybrid glycerohydrogels based on silicon tetraglycerolate and chitosan were produced by means of one-pot sol-gel synthesis. The influence of the component composition (the concentration of the template and precursor), the low-molecular-weight accelerator (NaCl), and the conditions of the process (pH, temperature) on the gelling duration was estimated. The glycerohydrogel surface microrelief was visualized by AFM in the semi-contact mode and the standard roughness parameters were calculated. The silicon-chitosan-containing glycerohydrogels were established to have a complex surface relief. Their solid sol-gel matrix is characterized by a uniform distribution of elements of the inorganic phase and a developed microporous structure with amorphous-crystalline ordering. The scale of the surface irregularities, as well as their structural and morphological features and the crystallinity degree of the solid sol-gel matrix, are determined by the template/precursor weight ratio and the molecular weight of the polymer.

1. Introduction

Inorganic/organic hybrid hydrogels are promising materials for application in medicine and pharmacology [1]. Such hydrogels are prepared by introducing an inorganic phase precursor (commonly silicon alkoxy compounds $Si(OR)_4$) into a polymer solution, resulting in the sol-gel synthesis of an inorganic network of siloxane bonds on an organic template matrix [2–4]. The possibility of one-pot synthesis of homogeneous monolithic hydrogels is an essential merit of sol-gel technology [5, 6].

The nature of the precursor and template, which determines the composition and structure of the hydrogel system [4, 6–8] and has a significant impact on the spectrum of its biologically useful properties [3, 4, 9–12], is the most important factor in obtaining hybrid hydrogels. E.g., the use of an ion-crosslinked biologically active chitosan-containing matrix as the organic phase allows obtaining biocompatible hydrogels with high mucoadhesive, bacteriostatic, and immunocorrecting properties [9, 10]. Using Si tetraglycerolates as the inorganic network precursor leads to the formation of silicon-chitosan-containing glycerohydrogels with haemostatic and wound-healing activity [3, 4, 6]. In this case, the precursor hydrolysis results in a spatial network of ≡Si–O–Si≡ bonds, and the chitosan macromolecules serve as a template during gelation [4, 6–8]. Besides, the inorganic phase formation proceeds non-catalytically under mild conditions (close to the physiological ones) to form a non-toxic by-product (glycerol), being a solvent of the resulting silicon polyolates; all this leads to the production of glycerohydrogels [11, 12].

In addition, chitosan aerogels, cryogels, and xerogels, formed from the corresponding hybrid hydrogels, are characterized by high porosity, large specific surface, satisfactory strength, and kinetic stability (not subject to syneresis) [4, 5]. This makes it possible to use chitosan-containing solid sol-gel matrices as bioadhesive tissue engineering designs (bioceramics, bio-glass) for the restoration of bone and cartilaginous tissues [13]. The presence of macro- and mesopores can be used to load drugs and growth factors; it also promotes faster tissue germination and filling of large-volume defects [14]. The dehydration method (sublimation or supercritical drying, freezing-thawing or extraction), as well as the formation conditions of the initial hydrogels, are the determining factors in the formation of a predetermined structure of the solid sol-gel matrix.

The surface roughness scale of hydrogel systems is known to determine the adhesion and proliferation of cells, and to be responsible for their dermoadhesive and mucoadhesive properties [15]. The solid sol-gel matrix surface topography plays an important role at the first stages of reparative osteogenesis. At present, atomic force microscopy (AFM) is the most promising direct method for analyzing the surface structure. This technique allows studying the surface geometry and local properties of a hydrogel sample with high spatial resolution and contrast [16, 17]. In addition, AFM studies in a semi-contact mode allow minimizing mechanical damage to the soft hydrogel surface. The subsequent digital data processing allows an in-depth analysis of the sample microrelief. To evaluate structural-morphological features of the solid sol-gel matrix, scanning electron microscopy (SEM), energy-dispersive X-ray spectroscopy (EDX), and X-ray analysis are considered as the most informative ones [18, 19].

The purpose of this work was to prepare silicon-chitosan-containing glycerohydrogels by sol-gel technology and to explore structural-morphological features of the surface of this glycerogohydrogel material and its solid phase using AFM, SEM, EDX, and X-ray diffraction analysis. A silicon tetraglycerolate solution in a two-molar glycerol excess ($Si(OGly)_4 \cdot 2$ GlyOH) and an ion-crosslinked hydrogels of chitosan glycolate were used as the precursor and template, respectively, for the preparation of hybrid glycerohydrogels. Preparing the ion-crosslinked chitosan matrix is based on the salt formation process occurring when chitosan is dissolved in an aqueous solution of glycolic acid [8]. The choice of this acid is due to the fact that it belongs to the class of pharmacopoeial substances (UNII: 0WT12SX38S), exhibits antioxidant properties, and promotes cell regeneration and moisture retention in tissues [20]. Due to these biologically useful properties, it is used to produce pharmaceutical, dermatological, and cosmetic preparations with keratolytic and rejuvenating properties for the treatment of acne, scars, and depigmentation [21].

2. Methodology

2.1. Materials

The following reagents were used: powders of chitosan with a viscosity-average molecular weight 38 (LCS) and 200 kDa (HCS), a deacetylation degree of 70 and 82 mol.%, respectively (Bioprogress Ltd., RF); tetraethoxysilane $Si(OEt)_4$ (Ekos-1 Ltd., RF); glycerol (GlyOH) (Vekton Ltd., RF); glycolic acid 70% $C_2H_4O_3$ (GlA) (Sigma-Aldrich, USA); NaCl and NaOH (NPO EKROS Ltd., RF); distilled water. All low-molecular-weight chemicals were of analytical grade or higher and used without further purification; freshly prepared solutions were always used in all experiments.

2.2. Synthesis of organically modified silicon

Silicon tetraglycerolates were synthesized by transesterification of tetraethoxysilane in a polyol excess without a catalyst according to Larchenco et al. [4]:

$$Si(OEt)_4 + 6\ GlyOH \leftrightarrow Si(OGly)_4 \cdot 2\ GlyOH + 4\ EtOH\uparrow.$$

The first step was esterification between tetraethoxysilane and glycerol within 70-80°C under constant stirring until the phase interface disappeared. At the second step, the formed free EtOH was distilled off from the reaction mixture at 80°C under atmospheric pressure. The completion of this stage was controlled by cooling to 73°C and by the volume of the EtOH distilled off as an azeotrope (~75% of the theoretical value). At the third step, EtOH was removed from the system at 140°C under

15 mm Hg for 3 h. The reaction completeness and the purity of the inorganic phase precursor were monitored using IR and 1H NMR spectroscopy, by no signals of >C=O and >C=C< within 1705–1685 cm^{-1} and those of –O–C_2H_5 within 1.1–2.3 ppm, respectively, and also by refractometry [22].

2.3. Preparation of chitosan solution

Chitosan (CS) solutions with a polymer concentration C_{CS} = 4% (w/v) in aqueous glycolic acid (GlA) with a concentration of C_{GlA} = 1.5% (v/v) were used. The choice of such polymer concentration was due to the fact that chitosan accelerated gelation in weakly acidic media, when C_{CS} > 1% (w/v) [6–8]. The solutions were prepared by dissolving polymer powder in an aqueous GlA solution at the ambient temperature without light for 24 h. As a result, chitosan glycolate (CS-GlA) was formed. The pH of the system was adjusted by adding a 0.4 M aqueous NaOH solution.

2.4. Sol-gel synthesis of silicon-chitosan-containing glycerohydrogels

To synthesize silicon-chitosan-containing glycerohydrogels, chitosan and $Si(OGly)_4 \cdot 2$ GlyOH solutions were used in several weight ratios (5:1 ÷ 1:13) with/without a low-molecular-weight accelerator (NaCl powder, 1 wt.%). The component concentrations were controlled gravimetrically on Ohaus Adventurer AR 1530 scales (the accuracy of weighing ± 0.002 g) and were expressed in wt.%. The mixed solutions were stirred carefully until homogeneity (1–2 min) and kept at 4±0.5, 20±2, or 37±0.5°C under atmospheric pressure for sol-gel synthesis. The gel point was fixed by the loss-flow time, using "the tube inverting method". The gel point corresponded to the instant of time, after which the system did not flow within 40 s after the tube had been turned over at a given temperature. The concentrations of chitosan and silicon in the obtained glycerohydrogels varied in the range of C_{CS} = 0.7–3.7 wt.% and C_{Si} = 0.4–2.4 wt.%, respectively. The template/precursor weight ratio (C_{CS}/C_{Si}) was used to express the component composition of each mixture.

2.5. Solid phase preparation

The solid phase (xerogel) was prepared by exhaustive cold extraction combined with cryotreatment and re-extraction. Samples were placed into EtOH (95.6%) for 30 days. The ethanol has been chosen, because it is a non-solvent (precipitant) for chitosan and silicon glycerohydrogel, but is a solvent for water, glycolic acid, and glycerol. The precipitant was replaced every 10 days. Then, the samples were cryofrozen in a refrigerator (Sanyo mdF-U3286S) at −85°C for 24 h and were re-kept in the precipitator (alcohol/acetone) for 3 days followed by drying in a desiccator ($CaCl_2$) during 24 h (Fig. 1).

2.6. Methods

The physicochemical parameters (pH and refractive index n_D^{25}) of the mixed solutions were measured on a Mettler Toledo Five Easy FE20 pH-meter and a Mettler Toledo RM40 refractometer (Germany).

IR spectra were recorded on a Nicolet IR-6700 spectrometer, FT-IR (USA) with a resolution of 4 cm^{-1} in the range of 4000 ÷ 5000 cm^{-1}, using a mean of 32 scans.

The 1H NMR spectra of the synthesized silicon tetraglycerols were recorded on a VARIAN-400 (USA) spectrophotometer within a frequency range of 5.4–2.0 ppm, 400 MHz. A solution of silicon tetraglycerolate (20 μl) in D_2O (480 μl) was used.

X-ray diffraction analysis was performed on a DRON-3 diffractometer with Cu-Kα emission at U = 22 kV and J_a = 20 mA. The crystallinity degree χ (%) was estimated by the method from Ref. [23] as the ratio of the total scattering from crystallites to the total scattering from both amorphous and crystalline regions:

$$\chi = [\int_0^\infty s^2\ I_c(s)ds/\int_0^\infty s^2\ I(s)ds]\cdot 100\%\ ,$$

where $s = (2\sin\Theta)/\lambda$, 2Θ is the diffraction angle (deg), λ is the X-ray wavelength equal to 1.54 Å; $I(s)$ and $I_c(s)$ are the intensities of coherent X-ray scattering from the entire sample and that from its crystalline region, respectively (rel. units).

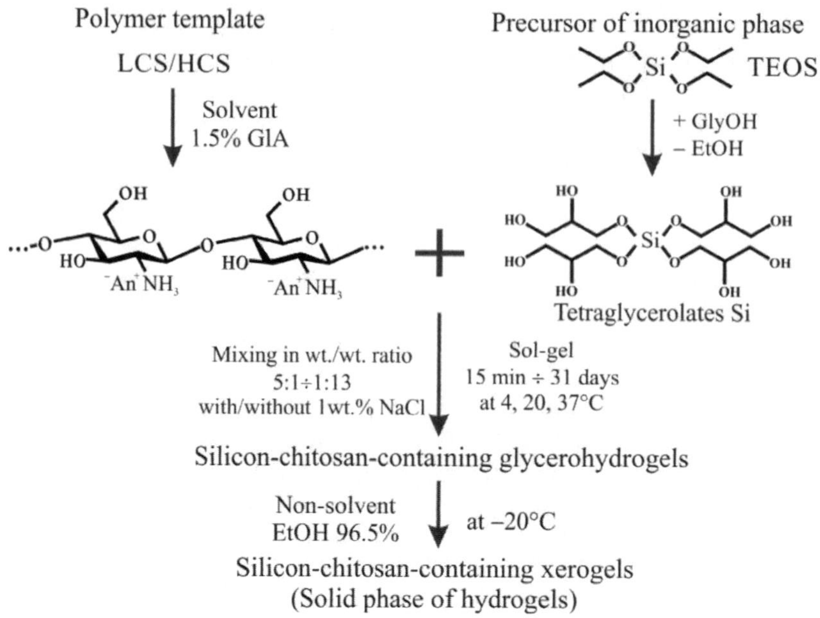

Figure 1. A scheme of the preparation of silicon-chitosan-containing glycerohydrogels and xerogels.

A graphical method was used to calculate χ (%). The X-ray diffractograms were digitized, the boundary between the amorphous halo and crystal peaks (the Paint editor) was drawn, and the areas of the resulting curvilinear figures were found (AreaS software).

The surface morphology and the Si content (C_{Si}*) in the surface layer of xerogel samples were evaluated by SEM and EDX on a MIRA/LMU scanning microscope (Tescan, Czech Republic), equipped with a INCA Energy 350 system (Tescan, Czech Republic) at a voltage of 8 kV and a conductive current of 60 pA. A 5-nm-thick golden layer was sprayed onto each sample with a magnetron sputtering installation K450X carbon coater (Germany) at a spraying current of 20 mA and a spraying duration of 1 min. The resolution was 3 nm, the INCA Energy detector sensitivity was 133 eV/10 mm^2. All studies were carried out in vacuum (10^{-2} Pa).

For image acquisition in a semi-contact mode, NSG10 probes (NT-MDT-SI Ltd., Russian Federation) with a resonance frequency around 220 kHz, a force constant 3–38 N×m^{-1} and a tip curvature less than 10 nm were used. AFM images of the surface of glycerohydrogels were processed by the *Gwyddion* 2.48 software [24] for the background flattening and for smoothing of three pixel length median filter to eliminate single-pixel noise emissions. Two of the standard roughness parameters: the mean roughness *Ra* and the root-mean-square roughness *Rq* were calculated.

3. Results and discussion

3.1. Study of the gelation time of silicon-chitosan-containing glycerohydrogels

To evaluate the effect of the component composition and the sol-gel synthesis conditions on the gelation time of silicon-chitosan-containing glycerohydrogels, mixed compositions with several template/precursor weight ratios (C_{CS}/C_{Si}) were prepared and held at three temperatures for the sol-gel process: the standard storage temperatures for pharmaceutical preparations (4 ± 0.5 and 20 ± 2°C) and the physiological temperature (37 ± 0.5°C). In addition, the pH of the mixed formulations was varied by adding NaOH to the CS-GlA stock solution. As the pH increased, the amount of CS deprotonated amino groups increased without precipitation of the polymer. To reduce the gelation time, a low-molecular-weight accelerator (NaCl) with the least toxic effect was chosen from the series of lyotropic anions [3]. In all cases, the formation of monolithic shape-stable transparent light-beige systems was completed within each sol-gel reaction (Fig. 2).

IOP Conf. Series: Materials Science and Engineering **256** (2017) 012011 doi:10.1088/1757-899X/256/1/012011

Figure 2. A sample of obtained silicon-chitosan-containing glycerohydrogel.

Figure 3 shows the dependence of the gelation time of the mixed compositions based on $Si(OGly)_4$ solutions in glycerol, low- and high-molecular-weight CS in GlA, on the template/precursor weight ratio for several pH values and temperatures. It can be seen that for all compositions in the whole studied concentration range of the polymer template and precursor, there is a general tendency of increasing the gelation time with an increase in the C_{CS}/C_{Si} ratio. Increasing the temperature and pH of the medium reduces the gelation time by ~1.5–12 times on average.

The electrolyte salt addition (NaCl) in an amount of 1 wt% further increased pH and reduced the gelation time of the mixed compositions (up to ~30 times) as compared to the initial composition with no accelerator. At the same time, the molecular weight of the polymeric template had no significant effect on time of reaching the gel point.

For more detailed study of the pH effect, a system based on LCS and $Si(OGly)_4 \cdot 2$ GlyOH with $C_{CS} = 2.67$ wt.% and $C_{Si} = 1.62$ wt.% was chosen, which gelation time in an acidic medium did not exceed 4 h. As expected, gelation time of this composition decreased monotonically with increasing pH (Table 1). E.g., in a strongly acidic medium (pH<4.2), gelation took several hours, while that in a slightly acidic medium (pH 4.3–5.0) took 30 min to 1 h. For pH 5.5–6.0 (close to neutral), gelation occurred within few minutes. Deceleration of the sol-gel transition in the lower pH range could be explained by the formation of a stable cationic complex of silicon glycerolate with the acid, where the silicon atom is in a coordinated saturated state [4, 6].

To evaluate the influence of the polymer template concentration on the gelling process, systems were chosen based on LCS and HCS with a constant precursor concentration ($C_{Si} = 1.62$ wt.%) and the same pH (4.5). Gelation was carried out at 37°C. As can be seen from Figure 4, both compositions are characterized by a reduced gelation time with increasing of the chitosan content. However, this effect is more pronounced for the HCS-based system than for the LCS-based one.

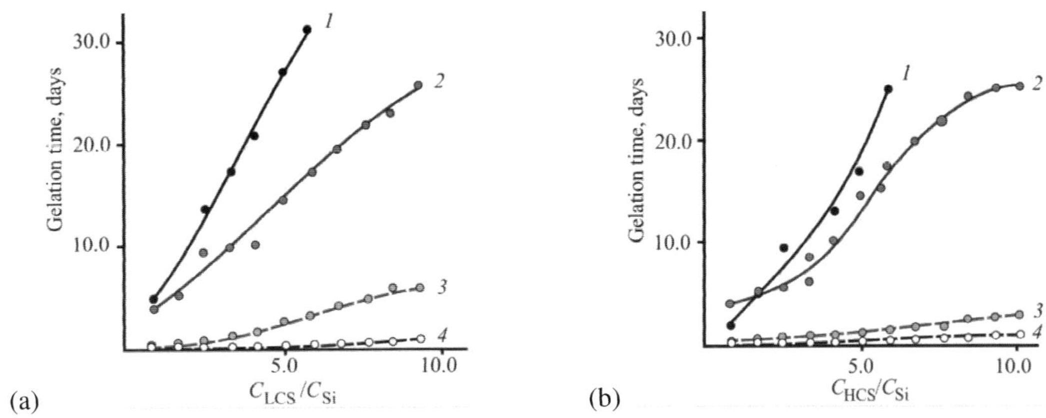

Figure 3. Dependence of the gelation time on C_{CS}/C_{Si} for the systems (a) LCS-GlA and (b) HCS-GlA + $Si(OGly)_4 \cdot 2$ GlyOH at several pHs: (a) 2.3 (*1*), 3.8 (*2, 3*), 4.3 (*4*) and (b) 3.3 (*1*), 3.8 (*2, 3*), 4.2 (*4*); the incubation temperature 4°C (*1, 2*), 20°C (*3*) and 37°C (*4*); without (*1-3*) and with addition of 1 wt.% NaCl (*4*).

Table 1. Dependence of the gelation time on pH for the system with C_{LCS} = 2.67 wt.% and C_{Si} = 1.62 wt.% at 37°C.

pH	Gelation time, h
3.8	3.6
4.0	2.3
4.2	2.0
4.7	0.8
5.0	0.5
5.5	0.3
5.7	0.2
6.0	0.1

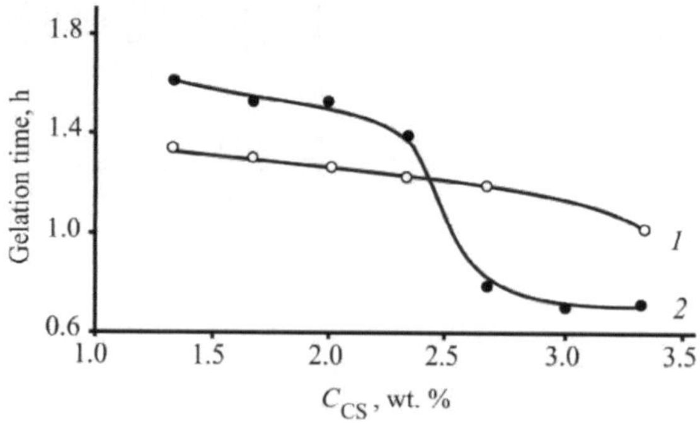

Figure 4. Dependence of the gelation time on C_{CS} for the system based on LCS (*1*), HCS (*2*) + $Si(OGly)_4 \cdot 2$ GlyOH at C_{Si} = 1.62 wt.%, 37°C, and pH = 4.5.

The dependence of the gelation time on the chitosan concentration (Fig. 4) correlates with the pH dependence (Table 1). The obtained dependence (Fig. 4) confirms once more the assumption from Ref. [6] that numerous intermolecular bonds are formed in a weakly acidic or neutral medium, including hydrogen bonding between the free (deprotonated) NH_2 groups of the template macromolecules and the Si-OH groups formed as a result of the precursor hydrolysis. The additional nucleation centers thus formed have an accelerating effect on the gelation. The template/precursor weight ratio in the mixture composition, which determines the hybrid glycerohydrogel morphology, is also important in this process.

Based on the obtained results, the following samples compositions were selected for the analysis of their structural-morphological features: with template/precursor weight ratios $C_{CS}>>C_{Si}$, $C_{CS}\approx C_{Si}$, and $C_{CS}<<C_{Si}$. The concentrations of the main components and the symbolic designations for the samples studied are given in Table 2.

3.2. AFM studies of the surface microrelief of glycerohydrogels

Topography visualization and a quantitative assessment of the surface microrelief of the obtained hybrid glycerohydrogels were performed by AFM. Analysis of the AFM images of surface areas (the scanning area size of 2×2 and 10×10 μm^2) showed that all samples were characterized by a complex surface relief (Figs. 5 and 6). Two types of structural irregularities were observed: protrusions and depressions (of spherical or irregular shape).

The alternation of these protrusions and valleys was fairly uniform. Sections were drawn on the AFM images, along which profiles of the surface roughness were plotted. The sizes of these irregularities (their height, depth, pitch, and diameter) and two roughness parameters (the arithmetic mean *Ra* and the mean-square roughness *Rq*) were estimated.

Table 2. Composition of the silicon-chitosan-containing hydrogel samples.

Sample	Chitosan	Composition, wt.%		Component weight ratio	Designation
		C_{CS}	C_{Si}	C_{CS}/C_{Si}	
1		3.3	0.8	4.1	LCS-4.1
2	LCS	2.0	2.4	0.8	LCS-0.8
3		0.7	4.1	0.2	LCS-0.2
4		3.3	0.8	4.1	HCS-4.1
5	HCS	2.0	2.4	0.8	HCS-0.8
6		0.7	4.1	0.2	HCS-0.2

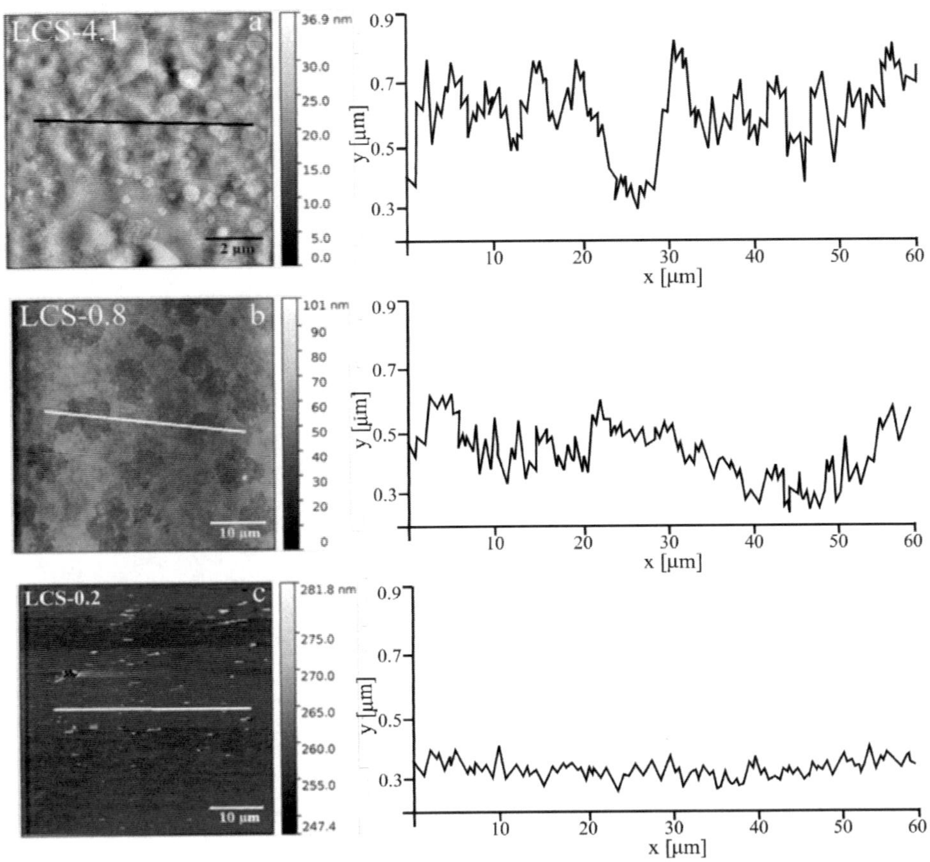

Figure 5. AFM image and roughness profile of the surface of LCS-based silicon-chitosan-containing glycerohydrogels. The designation of the samples is indicated in Table 2.

Figure 5 shows the AFM image and surface roughness profile of the obtained hybrid LCS-based glycerohydrogels. These samples are characterized by a granular relief with the predomination of protrusions of ~40–250 nm in height and a vertical pitch of ~75–500 nm. Statistical analysis showed that the mean arithmetic roughness of the surface varied within the range of Ra = 4–17 nm, and the mean-square roughness varied within the range of Rq = 5–22 nm (Table 3). The graininess and roughness scale is determined by the C_{LCS}/C_{Si} weight ratio: the surface irregularities naturally increase with the LCS content in the samples. The samples of our hybrid HCS-based and LCS-based glycerohydrogels are characterized by a granular relief, with the predominance of depressions of ~30–70 nm in depth and ~15–3 μm in diameter (Fig. 6). With decreasing the chitosan content in the sample, the diameter of the depressions decreased, and their surface concentration (the number of surface cavities per unit area) increased. This was the most characteristic for HCS-0.2. The roughness of these samples was also determined by the HCS content in the glycerohydrogel, but there was no clear pattern of the effect of the C_{HCS}/C_{Si} weight ratio on the micro-relief profile of the hydrogel surface. The mean arithmetic surface roughness of HCS-based samples was Ra = 10–32 nm, the mean-square roughness was Rq = 15–40 nm (Table 3).

3.3. Study of the solid phase (xerogels) of hybrid glycerohydrogels by EDX, SEM, and XRD

A quantitative analysis of the inorganic phase composition was carried out and structural-morphological features of the solid samples (xerogels) prepared from the corresponding glycerohydrogels by exhaustive cold extraction in combination with cryogenic processing and subsequent freeze drying were evaluated.

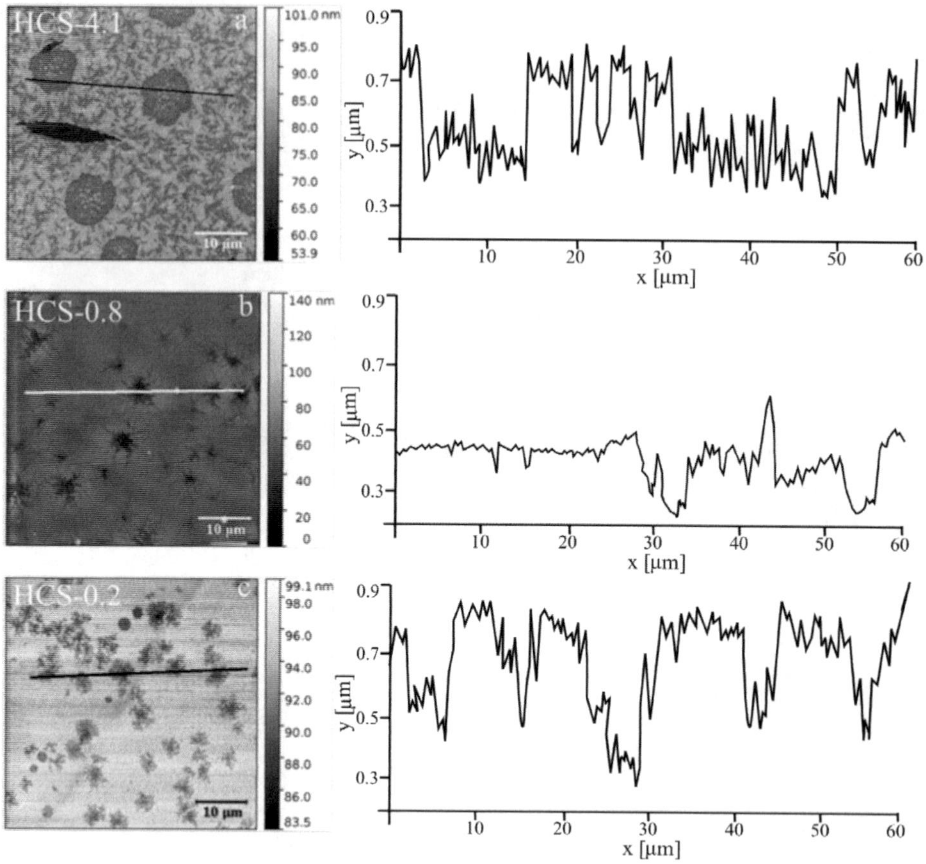

Figure 6. AFM image and roughness profile of the surface of HCS-based silicon-chitosan-containing glycerohydrogels. The designation of the samples as in Table 2.

The results of our quantitative estimation of the silicon content (C_{Si}*) in the surface layer are presented in Table 3 (HCS-based xerogels). Attention is drawn to the correlation of the C_{Si}* values in the solid samples with the calculated C_{Si} values in the corresponding glycerohydrogels (Table 2), which indicates a uniform distribution of the inorganic phase over the synthesized samples.

The SEM images show that the solid phase of the samples is represented by a developed microporous structure (Fig. 7). The samples with the highest chitosan contents (a, d) are characterized by a more loose structure with the largest number of aggregates formed from the surface, relatively uniform in the size of pores and channels. The greatest amount of these inhomogeneities was observed for the xerogel obtained from the glycerohydrogel based on LCS-4.1. As the polymer template content decreased, both in the system on the basis of LCS (a-c) and HCS (d-f), a tendency of leveling structural heterogeneities was observed. For the xerogel obtained from the LCS-0.2 sample with the lowest template content, an almost smooth surface (c) was observed. This difference, as well as smoothing of the relief with a decreased CS content, can be due to changes in the structure (e.g., compression) of the sample during the liquid phase removal. The lower the content of the structure-forming component (the template) in the glycerohydrogel sample, the higher the stresses arising during the xerogel preparation and the more likely the process of collapsing the initially highly porous structure.

The X-ray diffractograms of the xerogels are typical for amorphous crystalline polymers with a low crystallinity degree (Fig. 8).

Table 3. Physicochemical characteristics of the silicon-chitosan-containing glycerohydrogels and xerogels.

Sample	Hydrogels		Xerogels			
	Ra, nm	*Rq*, nm	2 Θ, deg	*d*, Å	*χ*, %	C_{Si}*, wt.%
	AFM-data		RSA-data			EDX-data
LCS-4.1	17	22	22.8	3.90	25.3	–
LCS-0.8	9	15	23.0	3.87	10.6	–
LCS-0.2	4	5	24.6	3.48	12.5	–
HCS-4.1	32	40	27.0	3.30	18.7	8.39
HCS-0.8	10	15	23.0	3.87	14.2	28.08
HCS-0.2	24	32	23.0	3.87	8.4	39.23

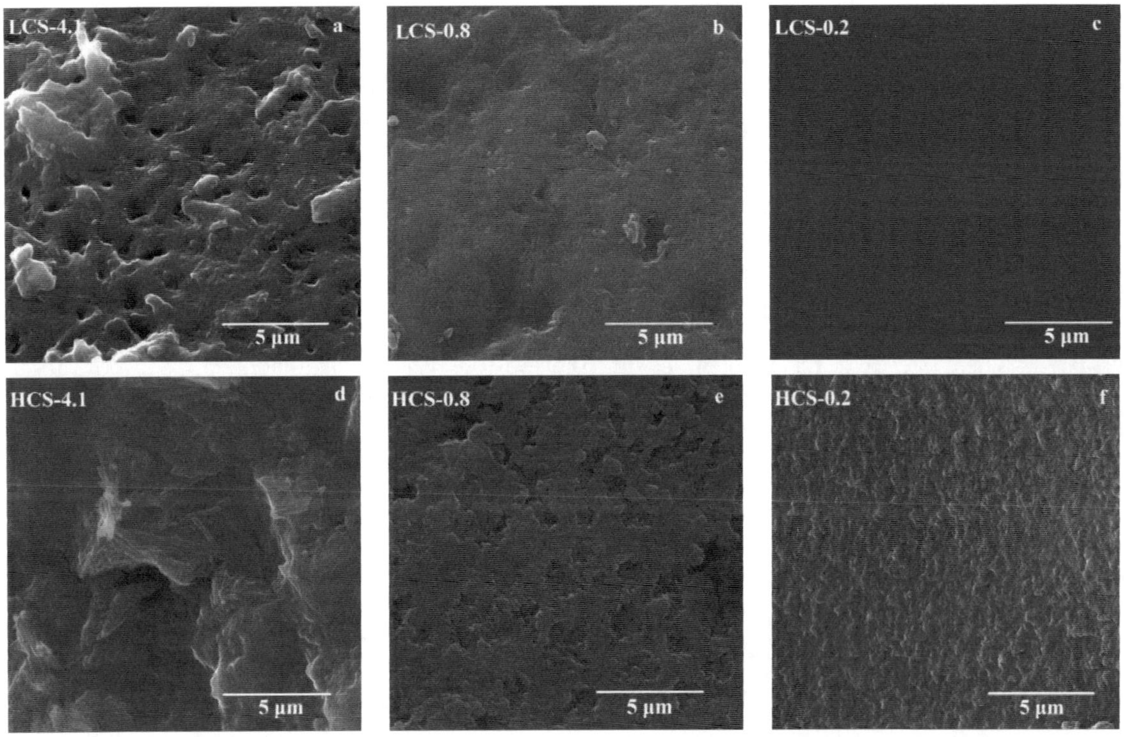

Figure 7. SEM images of the solid phase (xerogels) of the obtained silicon-chitosan-containing glycerohydrogels based on LCS (a-c) and HCS (d-f). The xerogel sample designation corresponds to that of the glycerogels, which they were obtained from (Table 2).

Against an amorphous halo background, reflections appear, characteristic of the crystalline structure of polymorphic chitosan at 2Θ = 22.8–24.6 deg (LCS) and 2Θ = 23.0–27.0 deg (HCS). The data on the interplanar spacings (*d*, Å) in the crystal lattices of the samples under study are close to each other (Table 3) and agree with the known literature data for chitosan films [25, 26].

The calculated crystallinity degree (*χ*) of solid samples decreased with increasing of the inorganic phase fraction (Table 3). For the LCS-based and HCS-based xerogels, the crystallinity degree was 25.3–10.6% and 18.7–8.4%, respectively. The lower crystallinity degree of the HCS-based xerogels could be explained by the more limited mobility of segments in the macromolecules of high-molecular-weight chitosan as compared to that with a lower molecular weight, which makes it difficult to form full-length long-range order.

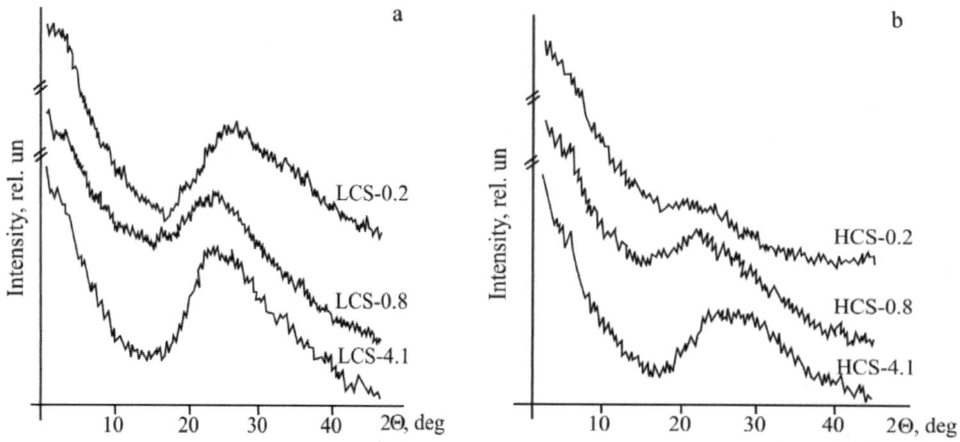

Figure 8. X-ray diffraction patterns of the xerogels of our silicon-chitosan-containing glycerohydrogels based on (a) LCS and (b) HCS. The xerogel sample designation corresponds to that of the glycerogels, which they were obtained from (Table 2).

4. Conclusion

Silicon-chitosan-containing glycerohydrogels based on HCS, LCS, and $Si(OGly)_4$ were prepared by sol-gel synthesis. The influence of the composition and the conditions of the process on the gelation time was evaluated. It was found that at pH values close to neutral, the gelation in the systems under study could proceed during 5–15 min.

Structural-morphological features of the solid phase of glycerohydrogels were visualized and evaluated. When $C_{CS}>>C_{Si}$, the xerogel specimens were found to be characterized by a more loose structure, higher values of their roughness parameters, and a higher crystallinity degree. When $C_{CS}<<C_{Si}$, a tendency was observed to smoothing out the heterogeneities and amorphization of the structure. Thus, variations of the composition of the template-precursor mixtures and the synthesis conditions allow obtaining silicon-chitosan-containing materials with the desired properties.

Acknowledgments

The results of this work were obtained within the framework of implementation of Grant «U.M.N.I.K.» No. 0032997 (Fund for Assistance to Small Innovative Enterprises in Science and Technology). The authors thank Tatyana G. Khonina and Elena V. Shadrina (Postovsky Institute of Organic Synthesis, Yekaterinburg, Russian Academy of Sciences) for their help in mastering the silicon glycerolate synthesis technique. The authors thank Vsevolod S. Atkin and Andrey M. Zakharevich (Nanomaterials and structures diagnostic laboratory of Education and Research Institute of Nanostructures and Biosystems of Saratov State University) for measuring the morphology and chemical elementary analysis of xerogel surface.

References

[1] Samal S, Dash M, Van Vlierberghe S and Kaplan D 2012 *Chem. Soc. Rev.* **41** 7147-94
[2] Shchipunov Y A, Karpenko T Y and Krekoten A V 2005 *Compos. Interfac.* **11** 587-07
[3] Larchenko E Y, Khonina T G, Shadrina E V, Pestov A V, Chupakhin O N, Menshutina N V and Chigvintsev S A 2014 *Russ. Chem. Bull.* **63** 1225-31
[4] Larchenko E Yu, Shadrina E V, Khonina T G and Chupakhin O N 2014 *Mendeleev Commun.* **24** 201-2
[5] Danks A E, Hall S R and Schnepp Z 2016 *Mater. Horiz.* **3** 91-112
[6] Shadrina E V, Malinkina O N, Khonina T G, Shipovskaya A B, FominaV I, Larchenko E Yu, Popova N A, Zyryanova I G and Larionov L P 2016 *Russ. Chem. Bull.* **64** 1633-9
[7] Malinkina O N, Sobolev A M and Shipovskaya A B 2016 *BioNanoSci* **6** 157-61

IOP Conf. Series: Materials Science and Engineering **256** (2017) 012011 doi:10.1088/1757-899X/256/1/012011

[8] Shipovskaya A B, Malinkina O N, Zhuravleva Yu Yu and Rogacheva S M 2016 Synthesis of silicon-containing chitosan hydrogels in a glycolic acid medium *Adv. Mater.Sci. Eng.* ID 3951703(2016)

[9] Bianchera A, Salomi E, Pezzanera M, Ruwet E, Bettini R and Elviri L 2014 Chitosan hydrogels for chondroitin sulphate controlled release: an analytical characterization *J. Anal. Methods Chem.* ID 808703(2014)

[10] Shipovskaya A B, Zudina I V, Fomina V I and Malinkina O 2015 *Butlerov Commun.* **41** 82-94

[11] Khonina T G, Chupakhin O N, Larionov L P, Boyakovskaya T G, Suvorov A L and Shadrina E V 2008 *Pharm. Chem. J.* **42** 609-13

[12] Khonina T G, Safronov A P, Shadrina E V, Ivanenko M V, Suvorova A I and Chupakhin O N 2012 *J. Colloid Interface Sci.* **365** 81-9

[13] Wang D, Romer F, Connell L, Walter C, Saiz E, Yue S and Jones J R 2015 *J. Mater. Chem.* B **3** 7560-76

[14] Lee E J, Jun S H, Kim H E, Kim H W, Koh Y H and Jang J H 2010 *J. Mater. Sci. Mater. Med.* **21** 207-14

[15] Drelich J and Mittal K L 2005 *Atomic force microscopy in adhesion studies* (Boston: RC Press)

[16] Oyen M L 2014 *Int. Mater. Rev.* **59** 44-59

[17] Drira Z and Yadavalli V K 2013 *J. Mech. Behav. Biomed.* **18** 20-8

[18] Jun S H, Lee E J, Yook S W, Kim H E, Kim H W and Koh Y H 2010 *Acta biomater* **6** 302-7

[19] Suryanarayana C and Norton M G 2013 *X-ray diffraction: a practical approach* (NewYork: Springer Science & Business Media)

[20] Ramos-e-Silva M, Hexsel D M, Rutowitsch M S and Zechmeister M 2001 *Clin. Dermatol.* **19** 460-6

[21] Kakudo N, Kushida S, Suzuki K and Kusumoto K 2013 *J. Cos. Derm.* **12** 281-6

[22] Kovyazin V A, Boev V V, Kopylov V M, Sokol'skaya I B 2008 *Russ. J. Gen. Chem.* **78** 216-22

[23] Cerqueira D A, Filho G R and Assuncao R M N 2006 *Polym. Bull.* **56** 475-84

[24] Nečas D and Klapetek P 2012 *Cent. Eur. J. Phys.* **10** 181-8

[25] Ageev E P, Vikhoreva G A, Zotkin M A, Matushkina N N, Gerasimov V I, Zezin S B and Obolonkova E S 2004 *Polym. Sci. Ser.* A **46** 1245-50

[26] Shipovskaya A B, Fomina V I, Kazmicheva O F, Rudenko D A and Malinkina O N 2017 *Polym. Sci. Ser.* A **59** 330-41

IOP Conf. Series: Materials Science and Engineering **256** (2017) 012012 doi:10.1088/1757-899X/256/1/012012

Local Young's moduli of as-grown and annealed diphenylalanine nanotubes

P S Zelenovskiy[1,*], V V Yuzhakov[1], S G Vasilev[1], A L Kholkin[1,2] and V Ya Shur[1]

[1] School of Natural Sciences and Mathematics, Ural Federal University, 620000 Ekaterinburg, Russia

[2] Department of Physics & CICECO – Aveiro Institute of Materials, University of Aveiro, 3810-193 Aveiro, Portugal

[*]zelenovskiy@urfu.ru

Abstract. Precise measurements of local transversal Young's moduli of as-grown and annealed nanotubes of diphenylalanine have been performed by mechanical nanoindentation. Evaporation of water from the nanochannels leads to pronounced reduction of the values' dispersion, and the characteristic value coincides with previous theoretical estimations. The experimental value of longitudinal Young's modulus measured at polar surface of vertically aligned bundle of nanotubes is reported for the first time.

1. Introduction

Self-assembly of bioorganic molecules into nanotubes is a convenient tool for manufacturing various functional micro- and nanodevices [1]. Recently, nanotubes of the simplest aromatic dipeptide diphenylalanine (D-Phe-D-Phe, FF) have attracted attention of researchers all over the world because of their pronounced piezoelectric [2, 3], ferroelectric [4, 5], and pyroelectric [6] properties. A number of new biosensors, biocompatible microelectronic and micromechanical devices based on FF nanotubes have been suggested [2, 7, 8].

Most of practical applications require the knowledge of FF nanotube mechanical properties, first of all the Young's moduli. However, these values, especially in a polar direction, are still poorly studied. Published values of transversal Young's modulus ranged from 4 up to 27 GPa [9-13]. This is definitely caused by small sizes of the nanotubes and unmanageable fixing conditions at their ends hampering the use of high resolution atomic force microscopy (AFM) [14, 15]. Moreover, water molecules remained in the nanotube's channel after the self-assembling and inhomogeneously distributed along the tube can drastically affect the obtained values [9].

In this work, we performed precise measurements of local transversal Young's modulus of as-grown (water filled) and annealed (empty) FF nanotubes by mechanical nanoindentation, which is devoid of the abovementioned shortcomings of AFM. Moreover, for the first time we measured longitudinal Young's modulus at the polar surface of a vertically aligned bundle of nanotubes.

2. Experimental details

FF microtubes representing bundles of individual nanotubes were grown on the metallized silicon substrate from alcohol-water solution during natural drying under ambient conditions (T=20°C, RH=35%) in accordance with a conventional technique [16]. A muffle furnace was used for thermal annealing of the microtubes. Annealing was performed in air at 80°C for 3 hours.

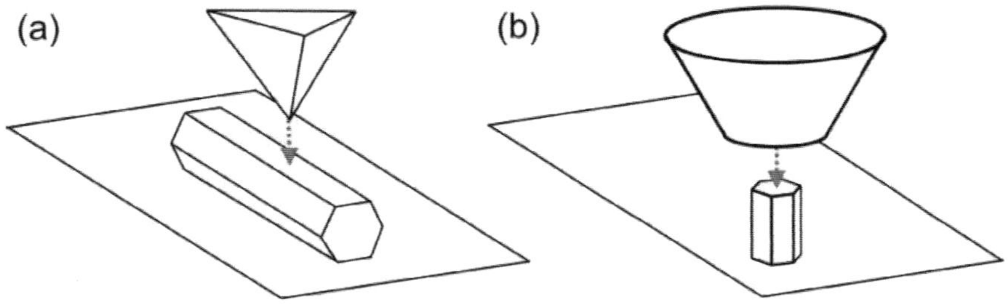

Figure 1. Schemes of the measurements of (a) transversal and (b) longitudinal Young's moduli by mechanical nanoindentation.

Local indentation measurements were performed using nano-hardness tester NanoScan-4D (FSBI TISNCM, Russia). The samples were fixed at a heavy aluminum cylinder. Measurements of transversal Young modulus were performed at nonpolar surface of the microtubes (Fig. 1a) using Berkovich indenter (diamond pyramid) with tip radius below 100 nm. Longitudinal Young modulus was determined by indentation at a polar cut of the microtube fixed vertically by the silver paste (Fig. 1b), using a "flat stamp" indenter with diameter about 100 μm. All measurements were carried out at ambient conditions.

3. Results and discussion

3.1. Transversal Young's modulus

Transversal Young's modulus (E_T) was measured at lateral (nonpolar) surface of the microtubes. As-grown tubes demonstrated pronounced bimodal distribution with two almost equivalent peaks. Characteristic values of these peaks are $E_{T1} = 9\pm3$ GPa and $E_{T2} = 24\pm5$ GPa (Fig. 2a). The first value, E_{T1}, is close to 8.75 GPa obtained in first-principles calculations [12], whereas the second value, E_{T2}, is close to 27±4 GPa obtained by AFM [10]. Such distribution has been attributed recently to water molecules remained in the tubes' channel and strongly affecting the value of E_T by the formation of radial network of hydrogen bonds [9]. The value E_{T1} was attributed to the Young's modulus of empty nanotubes, value E_{T2} – to Young's modulus of completely filled nanotubes, and intermediate values – to different amounts of water in the nanochannels [9].

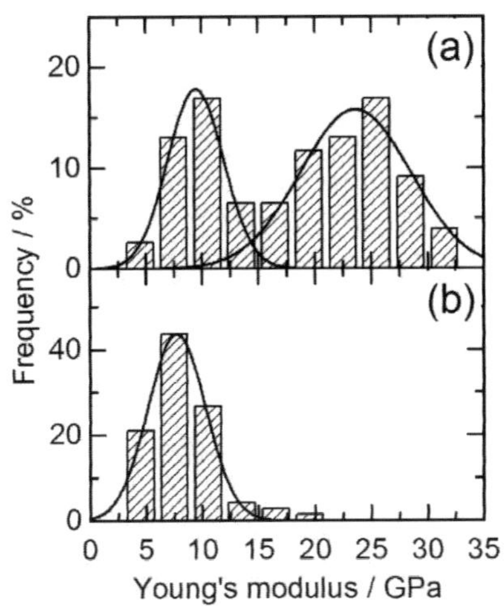

Figure 2. Distribution of transversal Young's modulus values of microtubes FF: (a) as-grown and (b) after annealing during 3 hours. Solid curves demonstrate Gaussian fitting.

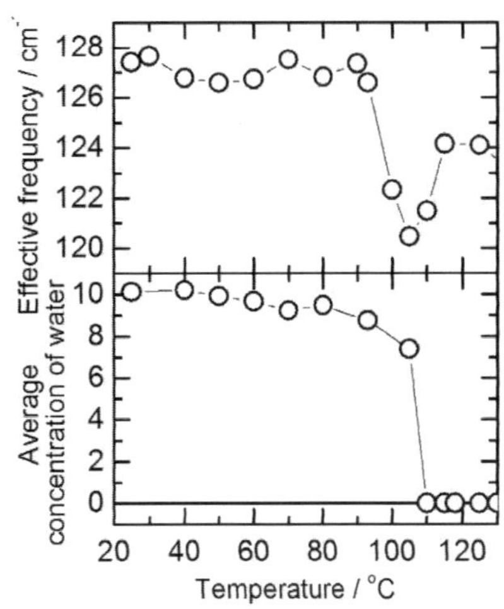

Figure 3. Temperature variation of effective frequency of nanotubes lattice vibration (top) and average concentration of water in the nanochannels (bottom).

IOP Conf. Series: Materials Science and Engineering **256** (2017) 012012 doi:10.1088/1757-899X/256/1/012012

To verify this hypothesis, we measured E_T of microtubes subjected to annealing for 3 hours at 80°C. Raman measurements performed earlier at various temperatures [17] showed that the effective frequency of nanotube lattice vibrations (and thus nanotube crystal structure) remained unchanged up to 105-110°C, whereas concentration of water in nanochannels (a number of water molecules per unit cell averaged over the volume about 0.025 μm^3, where Raman spectrum is measured) started to decrease gradually at 70-80°C (Fig. 3). Thus, the chosen annealing regime did not change the structure of the nanotubes, but promoted water evaporation from the nanochannels.

E_T distribution obtained for annealed microtubes demonstrated clear unimodal type with characteristic value 8±3 GPa (Fig. 2b). This value well coincides with that obtained in first-principles calculations for empty nanotubes [12], thus confirming the strong effect of water in the nanochannels on the mechanical properties of FF nanotubes suggested in [9].

3.2. Longitudinal Young's modulus

Measurements of longitudinal Young's modulus (E_L) along the polar axis of the nanotubes represent special interest. Such measurements are hampered by small size of the nanotubes and their preferable horizontal growth at the substrate. In our measurements, the bundle of nanotubes was manually separated from the substrate and fixed vertically by the silver paste (Fig. 1b).

It should be noted that conventional Berkovich-type indenter with tip radius below 100 nm can split the bundle into separate nanotubes during the measurements. Therefore, we used a "flat stamp" indenter with diameter of flat surface about 100 μm, which was much larger than diameter of the tube.

For estimation of the Young's modulus, we used conventional Hooke's law. In a case of uniform rod one can write $F = k \cdot \Delta l$, where F is a magnitude of the applied force, and Δl is a change of the rod's length. Spring constant k can be expressed via Young's modulus as $k = E \cdot S / l$, where area S of the tube section can be calculated from the optical image, and l is initial length of the tube. Therefore, for longitudinal Young's modulus we obtained $E_L = F \cdot l / S \cdot \Delta l$.

It is known that Hooke's law is valid for elastic deformations of the solids only. The value of elastic deformation, Δl, can be roughly estimated from the loading-unloading curves as a width of the unloading curve (Fig. 4). Mismatch between the widths of loading and unloading curves corresponds to plastic deformations. Since the plastic deformations always follow the regime of elastic ones, the value of the force F can be found from the loading curve as a value corresponding to Δl (Fig. 4).

The value of E_L obtained using the proposed method is 14±3 GPa. This is very close to 15.85 GPa obtained from first-principles calculations [12]. Theoretical estimations show that E_L weakly depends on the amount of water in the tube nanochannels, probably, because of high mobility of the molecules along the channels [9]. Therefore, the observed discrepancy is mainly due to overestimation of the value of elastic deformation, variation of the cross section area along the tube, and, thus, inexact determination of the force value. Nevertheless, for our best knowledge this experimentally obtained value is reported for the first time.

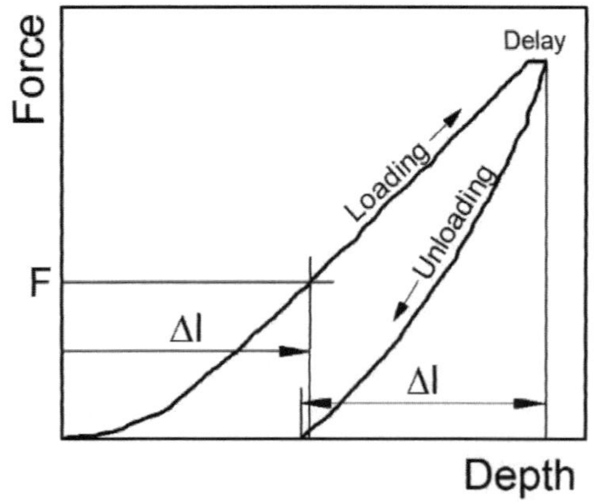

Figure 4. Scheme of elastic deformation and force value determination from loading-unloading curves for estimation of longitudinal Young's modulus.

4. Conclusion

The values of transversal Young's modulus in as-grown and annealed diphenylalanine nanotubes have been measured by mechanical nanoindentation. Systematic analysis of the results revealed the change of values distribution after annealing from bimodal to unimodal type with characteristic value 8 GPa. This confirms the earlier proposed hypotheses about the water effect on the Young's modulus of the nanotubes. For the first time, a value of longitudinal Young's modulus of about 14 GPa has been obtained experimentally. This value coincides with earlier reported first-principles calculations. Obtained results expand the understanding of mechanical properties of diphenylalanine nanotubes and will promote developing new functional devices on their base.

Acknowledgements

The research was carried out using equipment of UCSU "Modern Nanotechnologies" UrFU with the financial support by the grant of the President of the Russian Federation for young scientists (Contract 14.Y30.17.2294-MK) and the Government of the Russian Federation (Resolution 211, Contract 02.A03.21.0006).

References

[1] Lehn J-M 2002 *Science* **295** 2400-3
[2] Kholkin A, Amdursky N, Bdikin I, Gazit E and Rosenman G. 2010 *ACS Nano* **4** 610-4
[3] Vasilev S, Zelenovskiy P, Vasileva D, Nuraeva A, Shur V Ya and Kholkin A L 2016 *J. Phys. Chem. Solids* **93** 68-72
[4] Gan Z, Wu X, Zhu X and Shen J 2013 *Angew. Chem. Int. Ed. Engl.* **52** 2055-9
[5] Bdikin I, Bystrov V, Kopyl S, Lopes R P G, Delgadillo I, Gracio J, Mishina E, Sigov A and Kholkin A L 2012 *Appl. Phys. Lett.* **100** 043702
[6] Esin A, Baturin I, Nikitin T, Vasilev S, Salehli F, Shur V Ya and Kholkin A L 2016 *Appl. Phys. Lett.* **109** 142902
[7] Nguyen V, Zhu R, Jenkins K and Yang R 2016 *Nat. Commun.* **7** 13566
[8] Yemini M, Reches M, Rishpon J and Gazit E 2005 *Nano Lett.* **5** 183-6
[9] Zelenovskiy P, Kornev I, Vasilev S and Kholkin A 2016 *Phys. Chem. Chem. Phys.* **18** 29681-5
[10] Niu L, Chen X, Allen S and Tendler S J B 2007 *Langmuir* **23** 7443–6
[11] Kol N, Adler-Abramovich L, Barlam D, Shneck R Z, Gazit E and Rousso I 2005 *Nano Lett.* **5** 1343-6
[12] Azuri I, Adler-Abramovich L, Gazit E, Hod O and Kronik L 2014 *J. Am. Chem. Soc.* **136** 963-9
[13] Sinjab F, Bondakov G and Notinher I 2014 *Appl. Phys. Lett.* **104** 251905
[14] Chen Y, Dorgan B L, McIlroy D N and Aston E D 2006 *J. Appl. Phys.* **100** 104301
[15] Kluge D, Abraham F, Schmidt S, Schmidt H-W and Fery A 2010 *Langmuir* **26** 3020-3
[16] Nuraeva A, Vasilev S, Vasileva D, Zelenovskiy P, Chezganov D, Esin A, Kopyl S, Romanyuk K, Shur V Ya and Kholkin A L 2016 *Cryst. Growth Des.* **16** 1472-9
[17] Zelenovskiy P S, Davydov A O, Krylov A S and Kholkin A L 2017 *J. Raman Spectrosc.* Raman study of structural transformations in self-assembled diphenylalanine nanotubes at elevated temperatures On-line DOI: 10.1002/jrs.5084

IOP Conf. Series: Materials Science and Engineering **256** (2017) 012013 doi:10.1088/1757-899X/256/1/012013

High-speed precise cell patterning by pulsed electrohydrodynamic jet printing

A V Makaev*, E A Mingaliev, V R Karpov, I V Zubarev, V Ya Shur, and O S El'kina

Ural Federal University, Ekaterinburg, 620002, Russia

*makaev.andrew@yandex.ru

Abstract. The generation of micro-droplets of nutrient medium with living cells by pulsed electrohydrodynamic printing has been studied. In-situ visualization by high-speed camera made it possible to measure the characteristic times of droplet generation process and to determine the optimal printing parameters. Maximal frequency of stable generation was achieved at 700 Hz. This technique was applied successfully for drop-on-demand printing of culture medium with live HeLa cells and yeasts.

1. Introduction

Electrohydrodynamic (EHD) printing is a method of generating a droplet or a continuous jet by the application of an electric field to drop meniscus [1]. Among many direct writing methods, EHD printing has much potential for high resolution patterning [2]. EHD printing is driven by a strong electric field generated by a voltage between the nozzle with charged liquid and a grounded electrode [3, 4]. The essential diminishing of the droplet size as compared with the nozzle diameter allows producing micron-scale droplets without tube contamination [5].

Compatibility with different materials, applicability for large substrates, high targeting accuracy, relative simplicity, and non-vacuum and non-contact approach makes EHD printing attractive and suitable to wide range of applications. Moreover, the big viability (above 95%) [6] of cells subjected to EHD deposition on the substrate initiated its application in cellular and tissue engineering for the manipulation and placement of various cells on culture plates and scaffolds [7, 8].

Recently, in order to demonstrate drop-on-demand printing in the cone-jet mode, many researchers have adapted a square voltage pulse superposed by a dc bias voltage [9-13]. The short high-voltage pulse releases a droplet (or a finite number of droplets) from the nozzle, while the lower bias needs to ensure that a near conical shaped meniscus is always present, but not discharging any fluid [12, 13]. Therefore, the jetting start time can be shortened and the jetting frequency can be maximized [14, 15]. After termination of the rectangular voltage pulse, the meniscus may oscillate and recede back to the initial position through oscillation damping time [14]. Consequently, for a stable drop-on-demand jetting, the next pulse should be applied after the oscillation damping time. Besides, the meniscus oscillation may generate the unwanted secondary droplets. Therefore, to maximize drop-on-demand jetting frequency, it would be helpful to study the dynamics of the meniscus oscillation.

In this paper, we have demonstrated essential improvement of the main characteristics of EHD printing method using bias voltage with opposite sign to pulse voltage for minimizing the oscillation amplitude and the damping time. This technique was applied successfully for drop-on-demand printing of culture medium with live HeLa cells and yeasts.

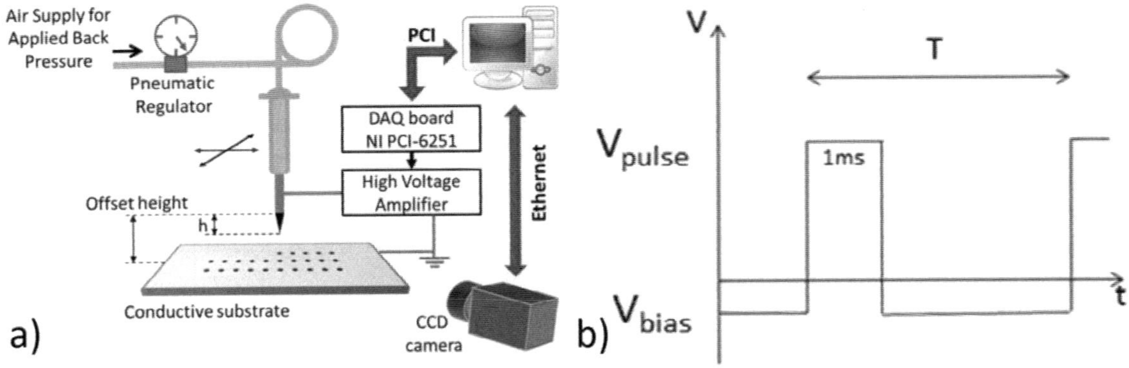

Figure 1. (a) Scheme of experimental setup for the droplet generation. (b) Waveform of applied voltage.

2. Experimental

The experimental setup consisted of a solution chamber fixed on three-axial translation stage, high-voltage power supply, and conductive substrate (Fig. 1a). The solution passed through a metal nozzle (inner diameter: 280 μm, outer diameter: 420 μm) at a constant flow rate controlled by air supply, which applied back pressure (≤ 5 psi) using pneumatic pressure controller (9THE-200, Taeha Corporation). The droplet generation was stimulated by the rectangular high voltage pulses (Fig. 1b) applied between the needle and glass substrate using the high-speed high-voltage power amplifier (Trek-10/10B-HS, TREK Inc.). The glass substrate was covered by 300-nm-thick Ti layer. The pulse amplitude below 2.5 kV and the current below 0.1 mA were used to prevent unwanted corona discharge and to ensure almost pure electrostatic force generation. Pulse duration equal to 1 ms was chosen to generate one droplet per pulse. The high speed CCD camera (Photron FASTCAM Mini UX100) with frame rate up to 10000 fps was used for in situ recording of the droplet generation and meniscus oscillation.

Multifunctional Cartesian robot with programmable motion controller (JC-3A00-0H3, Janome Sewing Machine CO. LTD.) was used for relocation and exact positioning of the solution camera in X-Y-Z directions. Such system allows depositing droplets according to the designed patterns. The arrays of dispensing droplets were visualized by optical microscope (Olympus BX-61) and scanning electron microscope (Auriga CrossBeam, Carl Zeiss).

Two types of solutions were used: (1) gelatin water solution and (2) DMEM medium. For gelatin solution, 50 mL of distilled water, 1 g of yeasts, and 1g of gelatin were added and the mixture was stirred for 30 min at 40°C. The solutions were kept at room temperature during 20 min and then were subjected to electrohydrodynamic printing. The HeLa cells were obtained from Institute of Immunology and Physiology UB RAS in Ekaterinburg. The cells were cultured in high-glucose Dulbecco's modified Eagle medium (H-DMEM; Invitrogen) with 10% fetal bovine serum (FBS; Hyclone) in a CO_2 incubator at 37°C and with 5% CO_2. The Hela cells were subcultured by trypsin (0.25%, invitrogen) dissociation at about 80% confluences.

3. Results and discussion

3.1. Visualization of droplet generation

EHD drop-on-demand printing exploits the high voltage pulses applied between the metal nozzle with liquid and the ground electrode, which induces charge accumulation at the liquid meniscus, forcing the meniscus to become more convex (i.e., increase the surface tension pressure) for balancing the electric pressure. The meniscus becomes conical and emits a fluid jet, when the electric pressure exceeds the surface tension pressure. In case of drop-on-demand printing, the constant V_{bias} forms near conical shaped meniscus, whereas the short-range V_{pulse} leads to jet emission. As noted above, the serious problem of printing is instability caused by oscillation of liquid meniscus after pulse voltage

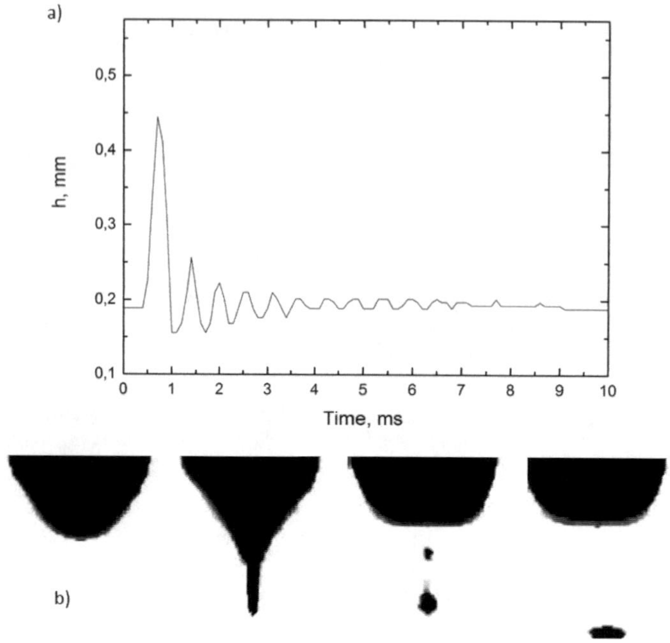

Figure 2. (a) The time dependence of the meniscus apex height h during droplet generation. (b) Main stages of the DMEM droplet generation.

termination leading to decrease of droplet generation frequency, and dripping of undesirable secondary droplets. The high-speed *in-situ* visualization of droplet meniscus behavior during generation process was carried out. The sequence of instantaneous images of drop meniscus is presented in Figure 2b.

The following main stages of the droplet generation were distinguished: (1) the deformation of the drop meniscus in the short-lived Taylor cone, (2) the emission of transient jet, (3) the dripping of a droplet, and (4) the damped oscillations of the meniscus. To characterize quantitatively the droplet generation process, the time dependence of the meniscus apex height h was extracted by specially developed software (Fig. 2a).

3.2. Generation of droplets at different bias voltages

To investigate the influence of V_{bias} magnitude on the damped oscillations, the time dependence of the meniscus apex height at different V_{bias} was calculated. The chosen V_{pulse} = 2.5 kV is slightly less than

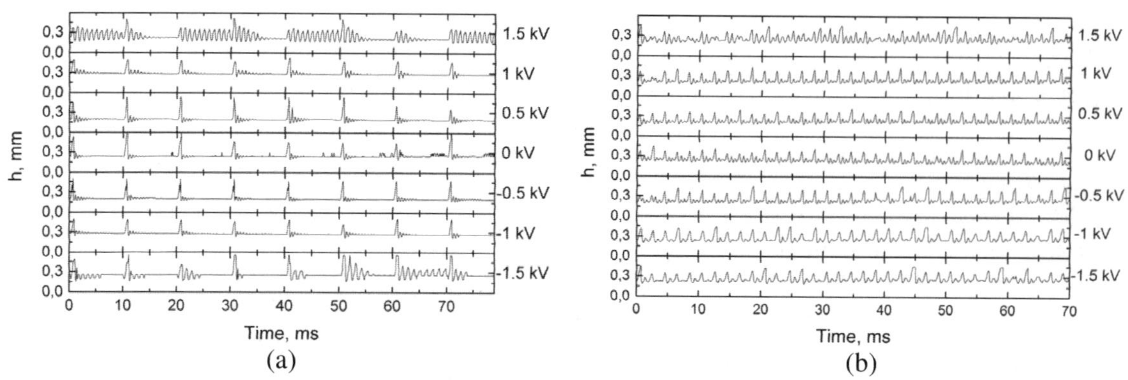

Figure 3. The time dependences of the cone apex height for pulse series with different V_{bias} for frequency: (a) 100 Hz, (b) 700 Hz.

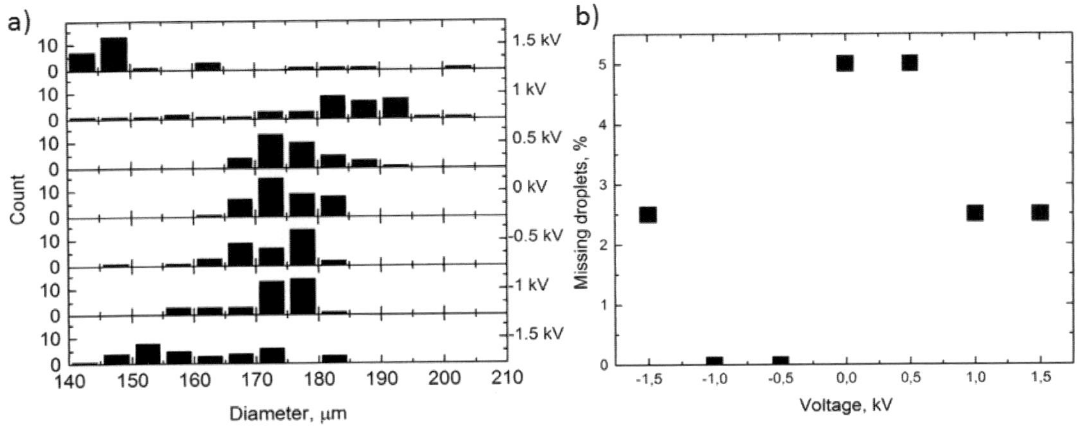

Figure 4. (a) Histogram of droplet diameter. (b) Percent of missing droplets in array.

the breakdown voltage leading to voltaic arc between nozzle and substrate. V_{bias} was varied between 1,5 kV and -1,5 kV at offset height of nozzle above the substrate 500 μm. Frequency of the pulse voltage generation was 100 Hz for nozzle translation speed v = 50 mm/s.

The jetting start time (time interval between the rising edge of the voltage pulse and the jet emission) was about 0.4 ms for all V_{bias}. The duration of transient jet existence was less than time resolution of the method, which was 0.1 ms. However, the duration of the damped oscillations exceeded several milliseconds and depended significantly on the sign and the value of V_{bias} (Fig. 3).

It was obtained that at high V_{bias} (+1.5 kV) the oscillation dumping time was comparable with pulse period (10 ms), being a consequence of the large amplitude of the oscillations after pulse termination (Fig. 3a). Moreover, we observed non-reproductive and unstable generation process – different lengths of transient jets and appearance of repeated jets in the interval between pulses leading to generation of unwanted secondary droplets. As a result, the droplets obtained at high V_{bias} have the wide range of diameters (Fig. 4a). It should be noted that there are 2.5% missing droplets in the array (Fig. 4b).

Decreasing of V_{bias} leads to reduction of both oscillation amplitude and damping time. As a result, the quantity of secondary droplets was reduced, but the missing droplets in the array were also present.

Then, we investigated the droplet generation using negative V_{bias}. As we can see in Figure 4, at V_{bias} = -1 kV and -0.5 kV the small variance of diameters and absence of missing droplets have been obtained. This can be explained by the braking force created by the counteracting electrical pressure, which effectively suppresses the meniscus oscillations after the pulse termination. Hence, shot oscillation dumping time and nonzero V_{bias} lead to the fact that at the beginning of the next voltage pulse the meniscus will take a conic shape with the apex in the previous position. Also, at these parameters the secondary droplets were not observed due to small amplitude of oscillations. At the same time, the dispensing accuracy (mean distance between center of droplets and programmed place) was slightly better (5.3 μm) than that for the best result obtained at the positive V_{bias} (8.8 μm).

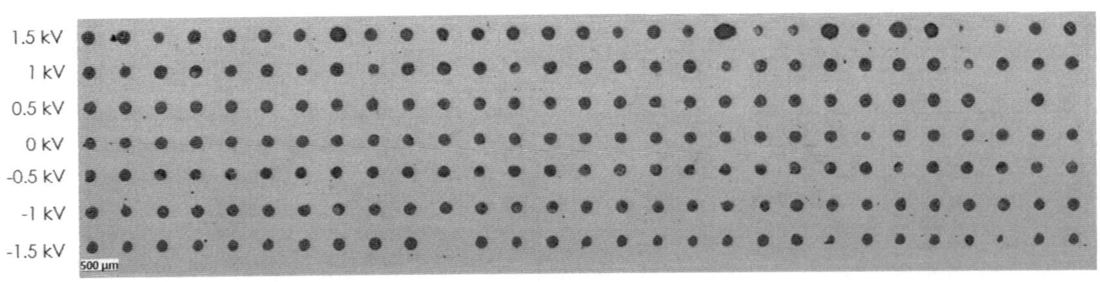

Figure 5. Optical images of linear arrays of DMEM droplets printed at different V_{bias}. V_{pulse} = 2.5 kV, f = 100 Hz, v = 50 mm/s.

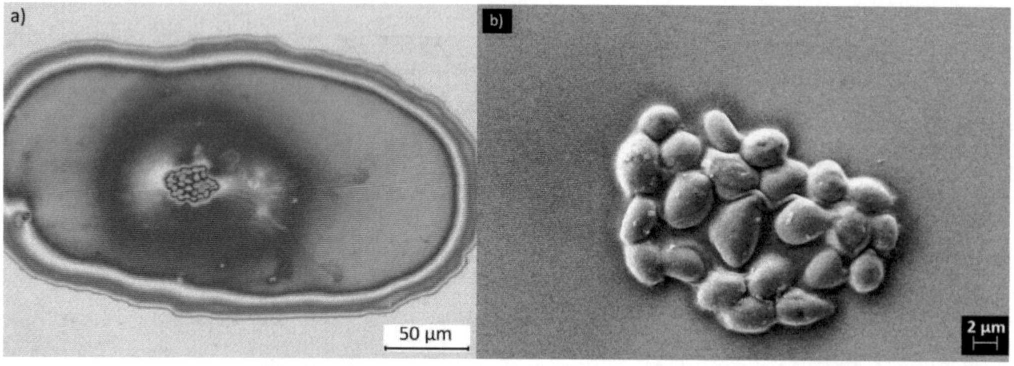

Figure 6. Gelatine droplets with few tens of yeast cells printed on glass substrate covered by 300-nm-thick Ti layer: (a) optical image, (b) scanning electron microscopy image.

Further increasing of negative V_{bias} worsened the results. As seen in Figure 3a, at V_{bias} = -1.5 kV continuous chaotic oscillations were observed. Electric pressure was too high at both directions and meniscus drop oscillations did not dump sometimes, being similar to behavior at V_{bias} = 1.5 kV (Fig.5).

At frequency of generation f = 700 Hz with nozzle translation speed v = 100 mm/s the difference in meniscus oscillations at various V_{bias} became smaller, but, nevertheless, small negative V_{bias} demonstrated the best accuracy and entirety of droplets array.

Summarizing, at high bias voltages (1.5 kV and -1.5 kV) we get unstable droplet generation with missing and secondary droplets. At lower bias voltages, we get rid of secondary droplets and dispensing droplets are obtained with small variance of diameters. But, at positive and zero voltages, percent of missing droplets in array increased. The best results without secondary and missing droplets and with narrow distribution of droplet diameter were obtained for generation at low negative V_{bias}. So, for suppressing meniscus oscillations we proposed to use negative voltage as a baseline voltage. For stable generation, it is necessary to ensure that a near conical shaped meniscus is always present, but not discharging any fluids. Therefore, negative voltage solves both problems – suppresses oscillations of meniscus and holds meniscus in the near conical shaped form.

3.3. Printing by living cells

This technique was used to realize drop-on-demand printing by culture medium with live cells. The regular arrays of gelatin solution droplets with yeast cells were printed on glass substrate covered with 300-nm-thick Ti layer in Figure 6. The few tens of cells were located in the center of each gelatine droplet in the region with diameter below 50 µm. Viability of yeasts was tested with trypan blue and was more than 95%.

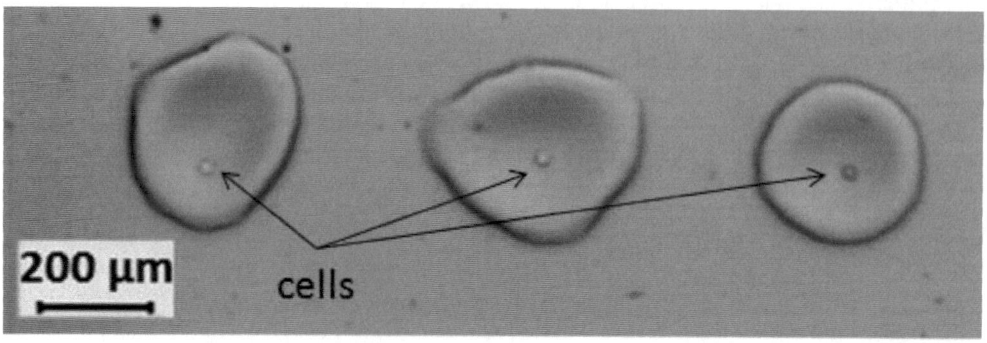

Figure 7. Droplets of Dulbecco's modified Eagle medium with HeLa cells.

The arrays of DMEM medium droplets with few HeLa cells (Fig. 7) in each were printed on the glass substrate covered with gelatin thin layer. The existence of the layer of gelatine, which was needed to prevent fast evaporation of the cultural medium, led to non-reproducibility of droplet sizes due to the ineffective screening of the induced static charges.

4. Conclusions

The high-speed droplet dispensing was achieved using electrohydrodynamic technique. Process of droplet dispensing was visualized with time resolution 0.1 ms and the main stages of droplet generation were distinguished. The measurement of the characteristic times of droplet generation process allowed defining parameters of voltage pulses for stable printing process and achieving maximum frequency of generation 700 Hz. The method was applied successfully for printing by culture medium with living cells: DMEM culture with HeLa cells and yeasts cell in gelatin water solution.

Acknowledgments

The research was carried out using the equipment of Ural Center for Shared Use "Modern Nanotechnologies" UrFU, with the financial support of the Government of the Russian Federation (Act 211, Agreement 02.A03.21.0006) and Ministry of Education and Science of the Russian Federation (Project No. 3.9534.2017/BP).

References

[1] Collins R T, Jones J J, Harris M T and Basaran O 2008 *Nat. Phys.* **4** 149–54
[2] Park J-U, Hardy M, Kang S J, Barton K, Adair K, Mukhopadhyay D K, Lee C Y, Strano M S, Alleyne A G, Georgiadis J G, Ferreira P M and Rogers J 2007 *Nat. Mater.* **6** 782–9
[3] Onses M S, Sutanto E, Ferreira P M, Alleyne A G and Rogers J A 2015 *Small* **11** 4237–66
[4] Taylor G 1964 *Proc. R. Soc. A Math. Phys. Eng. Sci* **280** 383–97
[5] Chen C H, Seville D A and Aksay I A 2006 *Appl. Phys. Lett* **89** 124103
[6] Zhao X, He J, Xu F, Liu Y and Li D 2016 *Virtual. Phys. Prototyp* **11** 57–63
[7] Gudapati H, Dey M and Ozbolat I 2016 *Biomaterials* **102** 20–42
[8] Murphy S V and Atala A 2014 *Nat. Biotechnol.* **32** 773–85
[9] Tran S B Q, Byun D, Nguyen V D and Kang T S 2009 *Phys. Rev.* E **80** 1–11
[10] Kwon K-S and Lee D-Y 2013 *J. Micromech. Microeng.* **23** 65018
[11] Kim J, Oh H and Kim S S 2008 *J. Aerosol Sci.* **39** 819–25
[12] Mishra S, Barton K L, Alleyne G, Ferreira P M and Rogers J 2010 *J. Micromech. Microeng.* **20** 95026
[13] Lee S, Song J, Kim H and Chung J 2012 *J. Aerosol Sci.*, vol. **52**, pp. 89–97
[14] Yang J, Kim H, Cho B and Chung J 2014 *J. Mech. Sci. Technol.* **28** 2815–23
[15] Li J L 2007 *J. Electrostat.* **65** 750–7

IOP Conf. Series: Materials Science and Engineering **256** (2017) 012014 doi:10.1088/1757-899X/256/1/012014

Study of structural colour of *Hebomoia glaucippe* butterfly wing scales

V Ya Shur*, D K Kuznetsov, V I Pryakhina, M S Kosobokov, I V Zubarev, S K Boymuradova, K V Volchetskaya

School of Natural Sciences and Mathematics, Ural Federal University, 620000 Ekaterinburg, Russia

*vladimir.shur@urfu.ru

Abstract. Structural colours of *Hebomoia glaucippe* butterfly wing scales have been studied experimentally using high resolution scanning electron microscopy. Visualization of scales structures and computer simulation allowed distinguishing correlation between nanostructures on the scales and their colour.

1. Introduction

The butterfly colour depends on the interaction of light with the material and spatial structure of wing scales. Three typical mechanisms of wing scales colouration are distinguished: (1) pigmentary colour, (2) structural colour, and (3) a combination of the two [1]. Pigments usually determine the colour of the scales. Additional colouration sometimes occurs because of the light interaction with complicated scale structures. The structural colours have been attributed to various physical mechanisms, such as multilayer interference, diffraction, Bragg scattering, and other [2]. The real colour of the butterfly wings can be understood by analysis of the inputs of the pigment absorption and light interaction with the structure of the wing scales.

In the present work, we investigate the origin of the different colourations of the wing scales of *Hebomoia glaucippe* butterfly both experimentally and by computer simulation.

2. Experimental methods

Samples of *Hebomoia glaucippe* butterfly (Fig. 1) have been received from Ekaterinburg Butterfly Garden, Russia. Butterfly wing scales were visualized using optical microscope (OM) Olympus BX51 (Olympus, Japan) and scanning electron microscope (SEM) with focused ion beam AURIGA™ Crossbeam Workstation (Carl Zeiss, Germany). For visualization of scales structure by SEM, the samples consisting of a butterfly wing or a part of the wing were mounted on a sample holder by a conductive carbon or a copper sticky tape. The samples were covered with 25-nm-thick Au/Pd thin film by magnetron sputtering to remove residual charge induced by electron beam. The scale structures were visualized using in-lens secondary electron detector at 5 kV electron accelerating energy. Cross-sections of wing scales were made by focused ion beam with liquid Ga emitter and ions accelerating energy 30 kV.

The reflectance spectra of butterfly wings were measured by spectrophotometer Agilent Cary 5000 (Agilent Technology Inc., USA) in the spectral range from 175 to 3300 nm. The samples were taped on microscope slide and placed perpendicular to a light path.

IOP Conf. Series: Materials Science and Engineering **256** (2017) 012014 doi:10.1088/1757-899X/256/1/012014

Figure 1. Hebomoia glaucippe butterfly.

3. Results

The wings of *Hebomoia glaucippe* have orange, white, and black colours (Fig. 1). The outer part of the dorsal forewings is bordered by narrow band of the black scales. Orange scales cover the main part of the dorsal forewings. White scales cover the rest of forewings and the whole hindwings.

SEM images of the wing scales from the dorsal side are shown in Figure 2. The studied scales of all colours have the similar sizes about 120x50 μm^2 and consist of ridges and crossribs. Three kinds of scales have been separated. The black and white scales have wavy ends (Fig. 2a,b), whereas the end of orange scale is rounded (Fig. 2c). Ridges extend longitudinally from one end of the scale to the other. The periods of ridges are 2.2 ± 0.1 μm for black scales, 2.3 ± 0.1 μm for white ones, and 1.1 ± 0.1 □m for orange ones. The periods of crossribs are 750 ± 50 nm and 1.2 ± 0.1 μm for black and white scales, respectively.

The structure of black and white scales can be attributed to unspecialized microstructure according to classification presented in [1]. The distinctive feature of the white scales is the presence of pigment beads that cover free space between ridges and crossribs (Fig. 2e).

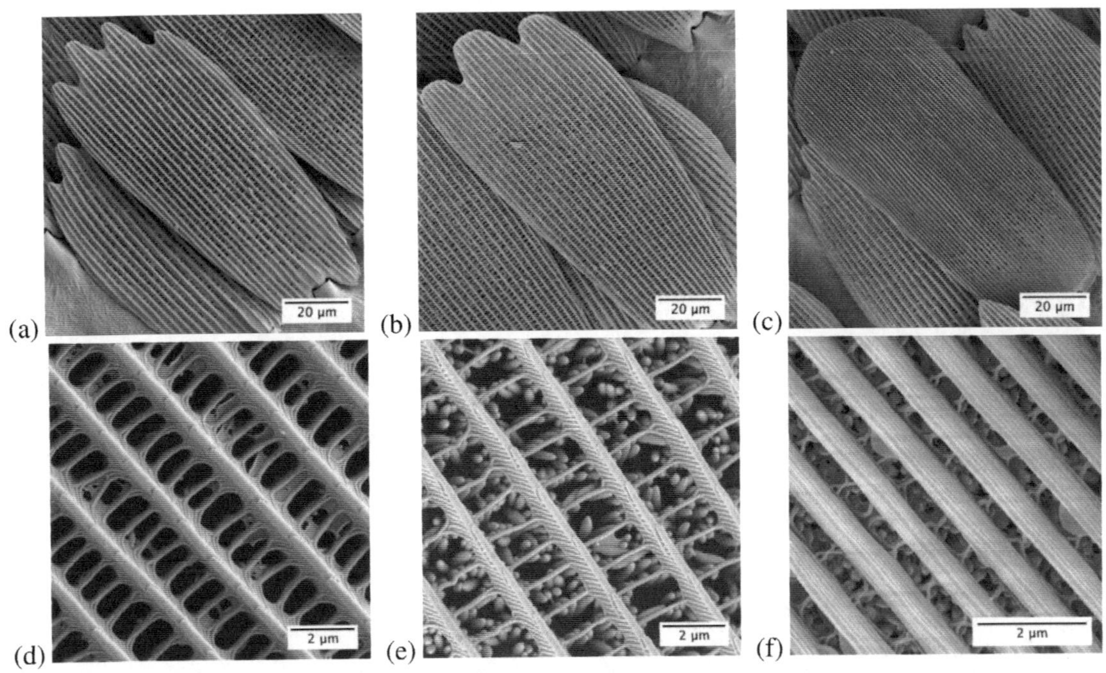

Figure 2. SEM images of the structures of *Hebomoia glaucippe* butterfly wing scales of different colours: (a,d) black; (b,e) white; (c,f) orange.

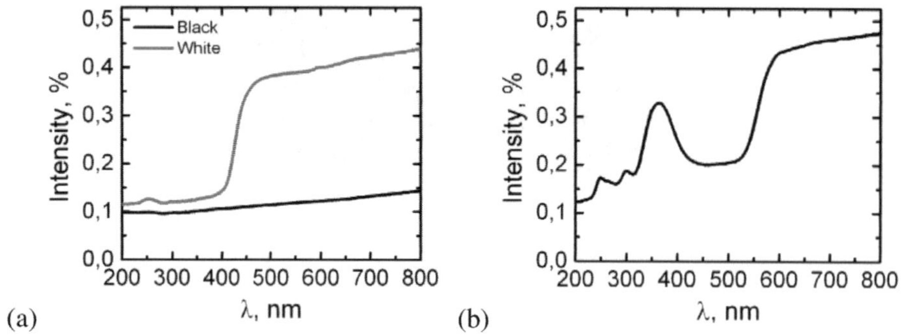

Figure 3. Measured reflectance spectra of the wing scales: (a) black and white colour; (b) orange colour.

The colouration of orange scales can be attributed to a ridge microstructure [1]. The ridges of these scales are folded into overlapped lamellae that run parallel to the scale surface. The ridge period is 1.1 ± 0.1 μm. The crossribs are very irregular and their period is difficult to estimate (Fig. 2f). The orange scales contain also the pigment beads but in essentially smaller portion than in the white ones.

The reflectance spectra were measured by spectrophotometer in spectral range from 200 to 800 nm. The black scales show very low reflectance in the whole spectral range (Fig. 3a), which can be attributed to the presence of melanin pigment contained in ridges and crossribs, which absorbs ultraviolet as well as visible light [3].

The white scales consist of ridges, crossribs, and pigment bead, which can play role of the scattering elements. Their reflectance is low for ultraviolet light (wavelength below 400 nm) and high for visible light (wavelengths above 450 nm) (Fig. 3). The low ultraviolet reflectance can be attributed to absorption by pigment located in the beads of the scales [3].

The reflectance spectrum from orange scales shows high intensity in the ultraviolet (from 300 to 450 nm) and in the visible wavelength range (above 550 nm) (Fig. 3b).

The computer simulation has been done for understanding, which elements of orange scales make contribution to reflection spectrum. The detailed information about ridge structure was obtained from its cross-section by focused ion beam. The platinum thin film was deposited on top surface of the scales to eliminate ridge damage by heavy ions during cutting. The ridge structure can be easily observed at the cross-section of the butterfly wing scale (Fig. 4a) due to its dark contrast in secondary electron image, as compared to lighter contrast of surrounding Pt film. It has been found that the ridge consists of six overlapped lamellae with thickness 55 ± 5 nm and period 255 ± 5 nm.

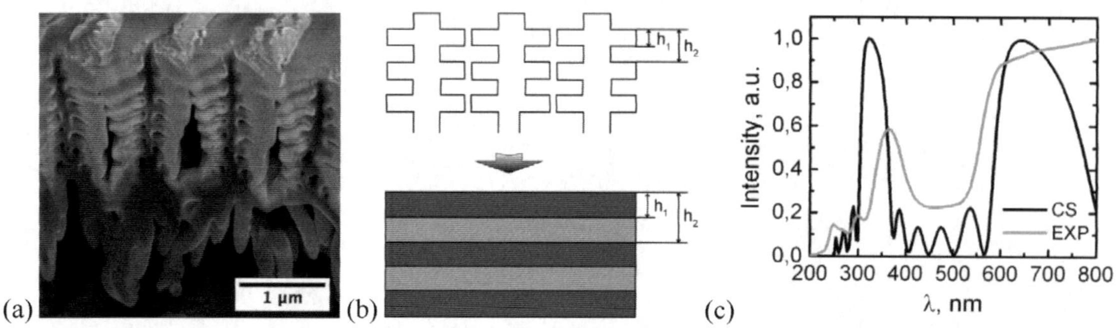

Figure 4. (a) Cross-section of the ridge scale obtained using focused ion beam cutting. (b) Multilayer model used for computer simulation of the reflectance spectrum. (c) Comparison of normalized measured (EXP) and simulated (CS) reflectance spectra.

During simulations, the parallel ridge lamellae structure was approximate with multilayer thin film structure (Fig. 4b). The layer thicknesses correspond to the respective values of lamellae. It was taken in mind that the ridges of butterfly wing scales consisted of chitin with refractive index about 1.5 [1], which was distinctly higher than the refractive index of air. The reflective spectrum was calculated using commercial finite-element modelling software COMSOL Multiphysics™. The obtained qualitative correlation of simulated spectrum with experimental one (Fig. 4c) allows concluding that the parallel ridge lamellae structure makes the main contribution to orange colour of the wing scales.

4. Conclusion

The microstructure of wing scales of the *Hebomoia glaucippe* butterfly was investigated. Three kinds of scales were found in butterfly dorsal wings. The cross-section of orange coloured scale allowed us to obtain the detail information about the ridge structure. Experimentally obtained parameters were used for computer simulation of the reflectance spectrum of orange scale using the two-dimensional planar multilayer model. The obtained qualitative correlation of simulated spectrum with experimental one allows concluding that the parallel ridge lamellae structure makes the main contribution to the orange colour of the wing scales.

The equipment of the Ural Center for Shared Use "Modern nanotechnology" SNSM UrFU was used. We thank Ekaterinburg Butterfly Garden (Russia) for granted butterfly samples. The work was supported by Government of the Russian Federation (Act 211, Agreement 02.A03.21.0006). D. Kuznetsov acknowledges financial support within the State Task from the Ministry of Education and Science of the Russian Federation (Project No. 3.7046.2017/7.8).

References

[1] Zhang D, Zhang W, Gu J, Fan T, Liu Q, Su H and Zhu S 2015 *Prog. Mat. Sc.* **68** 67
[2] Ghiradella H 1991 *Appl. Opt.* **30** 3492
[3] Wijnen B, Leertouwer H and Stavenga D 2007 *J. Ins. Phys.* **53** 1206

IOP Conf. Series: Materials Science and Engineering **256** (2017) 012015 doi:10.1088/1757-899X/256/1/012015

Combination of atomic force microscopy and mass spectrometry for the detection of target protein in the serum samples of children with autism spectrum disorders

A L Kaysheva[1,*], T O Pleshakova[1], A T Kopylov[1], I D Shumov[1], I Y Iourov[2], S G Vorsanova[2], Y B Yurov[2], V S Ziborov[3], A I Archakov[1] and Y D Ivanov[1]

[1]Institute of Biomedical Chemistry, Moscow, Russia.
[2]Mental Health Research Center, Moscow, Russia.
[3]Joint Institute for High Temperatures RAS, Moscow, Russia.

[*]kaysheva1@gmail.com

Abstract. Possibility of detection of target proteins associated with development of autistic disorders in children with use of combined atomic force microscopy and mass spectrometry (AFM/MS) method is demonstrated. The proposed method is based on the combination of affine enrichment of proteins from biological samples and visualization of these proteins by AFM and MS analysis with quantitative detection of target proteins.

1. Introduction

According to WHO, about 67 million people worldwide suffer from autism, and this level increases by 14% every year. Due to the complications in the diagnosis, in Russia there are no official data on the number of autistic children. In the early 2000s, the genes associated with the development of autism spectrum disorders (ASD) were annotated [1]. The genes associated with the development of ASD are, however, inhomogeneous and can participate in the development of other psychiatric and neurological disorders [2]. A full understanding of the causes of autism can be achieved by the revelation of functional protein markers along with behavioral reactions and the development of highly sensitive and efficient methods of their quantitative detection in biomaterial [3, 4].

Earlier, the authors carried out comparative panoramic mass spectrometric (MS) analysis of serum samples from three families, in which children with ASD were brought up [5]. As a result of the performed comparative analysis of the protein composition of the serum samples, a small group including 13 conventional marker proteins was sorted out; in this group, one *LIM domain-containing protein 1* was identified in four of five samples. To date, changes in the levels of GFAP, apoptosis factor Bcl-2, glutamate metabolism factor GAD-2, metallothioneins and thymidylate synthases, etc. [4, 6] in blood samples of children suffering from ASD are announced in scientific literature. These proteins are attributed to potential markers of development of autism in children.

The aim of the present study is the development of sensitive multiplexed method of target protein registration in serum samples of children suffering from ASD. The following objects were used as target proteins: apoptosis regulator (Bcl-2, UniProt AC P10415), metallothionein-3 (MT3, UniProt AC P25713), and thymidylate synthase (TYMS, UniProt AC P04818). The developed method is based on affine concentration of target proteins from the serum onto the surface of chips for atomic force microscope (AFM) with subsequent visualization of the affine complexes formed on this surface in a combination with MS quantitative analysis of the content of target proteins in biological samples [7].

2. Materials and methods

Similar to panoramic analysis [5], serum samples of children suffering from ASD were tested. Four control serum samples of healthy volunteers (C1, C4, C6, and C8) and five samples of children with ASD (S2, S3, S5, S7, and S9), provided by Separated Structural Unit "Clinical Research Institute of Pediatrics" at Pirogov Russian National Research Medical University Named After Y.E. Veltishev, were tested.

AFM chip represented an affine reagent, on which surface four sensor areas were formed: three working areas with immobilized monoclonal antibodies against three target proteins, and one control area without immobilized affine reagents. The procedures of AFM chips sensibilization are described in detail elsewhere [7]. AFM scanning of each area was carried out before and after the chip incubation in the analyzed sample [7].

Preparation of AFM chip for MS analysis included hydrolytic cleavage of proteins on the chip surface according to [7]. Mass spectrometric selected reaction monitoring (SRM) of the target proteins was carried out using Agilent 6490 mass spectrometer (USA), equipped with Agilent 1260/1290 HPLC system (USA) for unique peptides of target proteins according to the technique described in [7].

3. Results and discussion

AFM chips fabrication and analysis procedure (including the step of specific enrichment of AFM chip surface with target protein molecules) are described in our papers [5, 7, 8]. The identification of protein objects captured onto the AFM chip surface was carried out by SRM mass spectrometry [7].

The AFM scanning of working areas indicated that the affine complex «antibody/target protein» is hindered due to insufficient contrast of the images on the background of antibody molecules. That is, the changes in the heights of objects visualized in the working areas after the chip incubation in the target protein solution are insufficient for unambiguous evaluation of complex formation. Earlier, with HIV-1 gp120 glycoprotein we demonstrated that the best contrast of AFM images is observed for small probe molecules – aptamers (short single-stranded DNA sequences specific against the target protein). The contrast of «aptamer/target protein» images is twice higher than that of «antibody/target protein» images [9]. For this reason, in our present research AFM chip served as efficient affine reagent [5, 8]. Subsequent SRM analysis of the AFM chips has revealed the presence of Bcl2 protein on the surface of three chips and TYMS protein on the surface of one chip at picomolar concentrations (Table 1). Identification of MT3 protein on the AFM chips' surface was not possible.

It is interesting to point out that AFM/SRM method allowed detection of Bcl-2 protein in serum samples of brothers (S2 and S3), and also in S5 sample.

Table 1. Results of AFM/SRM analysis of 5 serum samples of children suffering from ASD, and of 4 control samples.

Sample No.	Protein	Results of AFM/SRM analysis	
		Amino acid sequence of the revealed peptide	Concentration of the revealed peptide ($\times 10^{-12}$ M)
S2	Bcl-2	FATVVEELFR	2.10
	TYMS	–	–
S3	Bcl-2	FATVVEELFR	4.20
	TYMS	DFLDSLGFSTR	20.8
S5	Bcl-2	FATVVEELFR	1.20
	TYMS	–	–

4. Conclusion
Thus, we have demonstrated the efficiency of SRM with preliminary affine enrichment and concentration of target protein molecules from serum onto small area of AFM chip surface for target protein detection.

Acknowledgements
This study was supported by Russian Science Foundation (Grant No. 14-25-00132). Shumov I.D. is a recipient of Russian Federation President scholarship for young scientists for 2016-2018 (project SP-4280.2016.4).

References
[1] Freitag C, Staal W, Klauck S, Duketis E and Waltes R 2010 *Psychiatry* **19** 169-78
[2] Guilmatre A, Dubourg C, Mosca A, Legallic S, Goldenberg A et al. 2009 *Arch. Gen. Psychiatry* **66** 947-56
[3] Hartley S, Seltzer M, Hong J, Greenberg J, Smith L, Almeida D, Coe Ch and Abbeduto J 2012 *Int. J. Behav. Dev.* **36** 53-61
[4] Khramova T, Kaysheva A, Ivanov Yu, Pleshakova T, Iourov I, Vorsanova S, Yurov Y, Schetkin A and Archakov A 2017 *J. Mol. Neurosci.* **62** 420-9
[5] Kaysheva A, Kopylov A, Yurov I, Vorsanova S, Yurov Yu, Galiullin R, Anashkina A, Archakov A and Ivanov Yu 2016 *Clinical Practice in Pediatrics* **11** 12-7
[6] Woods A, Sokolowska I, Taurines R, Gerlach M, Dudley E, Thorne J and Darie C. 2012 *J. Cell. Mol. Med.* **16** 1184-95
[7] Pleshakova T, Malsagova K, Kaysheva A, Kopylov A, Tatur V, Ziborov V, Kanashenko S, Galiullin R and Ivanov Yu 2017 *FEBS Open Bio* **7** 1186-95
[8] Kaysheva A, Ivanov Iu, Zgoda V, Frantsuzov P, Pleshakova T, Krokhin N, Ziborov V and Archakov A 2010 *Biochemistry (Moscow) Suppl. Series B: Biomedical Chemistry* **4** 15-24
[9] Ivanov Y, Bukharina N, Pleshakova T, Frantsuzov P, Andreeva E, et al. 2014 *Int. J. Nanomedicine* **3** 4659-70

IOP Conf. Series: Materials Science and Engineering **256** (2017) 012016 doi:10.1088/1757-899X/256/1/012016

Physical-mechanical image of the cell surface on the base of AFM data in contact mode

M N Starodubtseva[1,3,*], I E Starodubtsev[1,4], N I Yegorenkov[1], N S Kuzhel[2], E E Konstantinova[2] and S A Chizhik[2]

[1]Gomel State Medical University, Gomel, Belarus
[2]A V Luikov Heat and Mass Transfer Institute of NAS of Belarus, Minsk, Belarus
[3]Radiobiology Institute of NAS of Belarus, Gomel, Belarus
[4]Belarusian State University, Minsk, Belarus

[*]marysta@mail.ru

Abstract. Physical and mechanical properties of the cell surface are well-known markers of a cell state. The complex of the parameters characterizing the cell surface properties, such as the elastic modulus (E), the parameters of adhesive (F_a), and friction (F_f) forces can be measured using atomic force microscope (AFM) in a contact mode and form namely the physical-mechanical image of the cell surface that is a fundamental element of the cell mechanical phenotype. The paper aims at forming the physical-mechanical images of the surface of two types of glutaraldehyde-fixed cancerous cells (human epithelial cells of larynx carcinoma, HEp-2c cells, and breast adenocarcinoma, MCF-7 cells) based on the data obtained by AFM in air and revealing the basic difference between them. The average values of friction, elastic and adhesive forces, and the roughness of lateral force maps, as well as dependence of the fractal dimension of lateral force maps on Z-scale factor have been studied. We have revealed that the response of microscale areas of the HEp-2c cell surface having numerous microvilli to external mechanical forces is less expressed and more homogeneous in comparison with the response of MCF-7 cell surface.

1. Introduction

Physical and mechanical properties of the cell surface representing the layered composite material consisting of glycocalyx, lipid bilayer, and cytoplasmic layer with cortical cytoskeleton are well-known markers of a cell state. Basic cellular element contributing the mechanical properties of cell surface is the cortical cytoskeleton as a cell bearing support. The complex of the parameters characterizing such cell surface properties as the elastic modulus (E), adhesive (F_a) and friction (F_f) forces can be measured using atomic force microscope (AFM) in a contact mode. This complex of AFM parameters forms namely the physical-mechanical image of the cell surface that is a fundamental element of the cell mechanical phenotype [1]. The mechanical phenotype ("mechanotype") of cells is considered by many researchers to be a potential biomarker for cell types ranging from pluripotent stem to cancer cells [2, 3]. The complex of the parameters forming the physical-mechanical image of the cell surface includes not only the averaged characteristics of elastic, adhesive, and friction properties, but also the spatial distribution of these properties over the cell surface.

The paper aims at forming the surface physical-mechanical images of two types of cancerous cells (human epithelial cells of larynx carcinoma, HEp-2c cells, and breast adenocarcinoma, MCF-7 cells) based on the data obtained by AFM and revealing the basic difference between them.

2. Materials and methods

2.1. Preparation of cell samples for AFM

Samples of human primary skin fibroblasts, larynx carcinoma cells (HEp-2c cell line), and breast adenocarcinoma cells (MCF-7 cell line) were kindly provided by D. R. Petrenyov (Institute of Radiobiology, NAS of Belarus, Gomel, Belarus). The cells suspensions were placed on the specially prepared glass slides in Petri dishes with growth medium and incubated during 24 h at 37 °C and 5 % CO_2. Then the cells were fixed with 0.5 % glutaraldehyde (30 min) and washed with deionized water. The cells were dried in tilted position (75-85°) in air laminar stream (0.42 m/s) at room temperature.

2.2. Atomic force microscopy

The scanning was performed with an AFM ("NT-206", MicroTestMachines Co., Belarus) using standard probes (MikroMasch) in air under room conditions. CSC38 AFM cantilever type (lever B, nominal force constant of 0.03 N/m) was used. Topographic images and lateral force maps were recorded. Lateral forces determined mostly by the friction forces between the AFM probe tip and surface were assessed by measuring the cantilever's torsion value and represented in arbitrary units. Using the lateral force maps, the statistical parameters of the surface roughness (R_q) were assessed. The microscale AFM images of the cell surface without apparent defects were used for analysis. The cell surface regions for AFM analysis were chosen between the nucleus and peripheral region. The elastic modulus (Young's modulus) and adhesive force were assessed by force spectroscopy (force-curve analysis) [4] using blunted NSC11 (MikroMasch) cantilever tips (radius 76-81 nm, force constant 3.0 N/m). The indentation depth was 10 nm.

2.3. Fractal dimension estimation

Fractal dimension (D_F) was calculated for lateral force maps of the cell surface (2.5×2.5 μm²) using the following formula (box counting dimension):

$$D_F = \lim_{\varepsilon \to 0} \frac{\lg N(\varepsilon)}{\lg \frac{1}{\varepsilon}} \quad (1),$$

where N(ε) is the number of cubes with edge ε that together included the studied surface.

The area with the studied surface was divided by cubic lattice with cube edge length ε that was initially set as half of the area size. Then the number of cubes that included at least one point of the surface was calculated. After that, cube edge length ε was reduced by half and the process was repeated in loop until cube edge length became less than a certain constant depending on the AFM scanning step. The fractal dimension of the surface was estimated as a maximal slope coefficient of the obtained dependence between lg(N(ε)) and lg(1/ε). In the used modified box counting algorithm, the surface was initially divided into 8 equal fragments. After calculation of the dimension for each fragment, the fractal dimension of the whole surface was calculated as the mean value and limits of 95% confidence interval. To analyze the dependence of D_F on the Z-scale factor, X- and Y-data of the AFM image were not changed but Z-data was multiplied by t that was changed in a broad range. The fractal dimension was calculated using the box counting algorithm for each t value and the dependence D_F (t) was plotted and analyzed.

2.4. Statistical analysis

The correlation of data to normal distribution was checked with the Shapiro-Wilk's W test. Results are represented as limits of 95% confidence interval (CI). The sample means were compared with Student's t test.

3. Results and discussion

The basic morphological parameters of the studied cells (human fibroblasts and cancerous epithelial cells of lines: HEp-2c and MCF-7) are represented in Table 1. Figure 1 shows the typical lateral force maps of the cell surface.

(a) (b) (c)
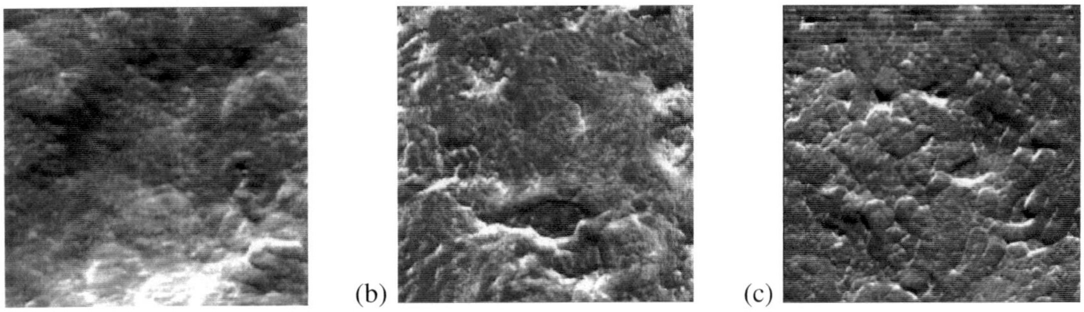

Figure 1. Typical lateral force images of cell surface areas in the middle part of (a) fibroblast, (b) MCF-7 cell and (c) HEp-2c cell. Image area: 2.5 × 2.5 μm^2.

Table 1. Morphological parameters of cells.

Cells	D_1, μm	D_2, μm	h, μm	D_2/D_1	D_1/h	D_2/h
Fibroblasts	60.4±26.6	22.3±6.9	1.5±0.3	0.39±0.15	40.7±12.8	16.3±8.7
MCF-7 cells	33.1±11.6	14.9±1.8	1.5±0.4	0.51±0.18	22.6±6.5	10.5±2.6
HEp-2c cells	28.1±2.4	11.1±2.5	1.0±0.3	0.40±0.08	30.0±9.0	12.0±4.1

D_1 and D_2 are the maximal and minimal sizes of cells measured in orthogonal directions in the plane of the cell contact to the glass surface; h is the maximal value of the cell height; D_2/D_1 is a parameter characterizing the cell polarization; D_1/h and D_2/h are parameters characterizing the cell spreading on a solid surface. Data are represented as the limits of 95% CI. n=5-13.

Though the cell sizes differ for the different cell types, especially in the case of fibroblasts and cancerous epithelial cells, nearly all the parameters showing the cell ability to spreading and polarization are the same.

To form the physical-mechanical images of the surface of the studied cells we used the complex of AFM parameters obtained in a contact mode. The complex included the statistical parameters (the mean and standard deviation) of the elastic modulus (Young's modulus, E), adhesive force (F_a) measured in different locations of the cell surface, the roughness (R_q) of lateral (friction) forces assessed over microscale lateral force maps of the cell surface, the fractal dimension (D_F) of microscale lateral force maps of the cell surface and its dependence on Z-scale factor (t). Parameter D_F=f(t) characterizes the structure of the distribution of the mechanical properties of the cell surface more comprehensive than a single value of the fractal dimension calculated for the AFM image of the cell surface as well as the roughness of the same AFM image [5].

Averaged values of the Young's modulus and adhesive force for two types of cancerous epithelial cells were significantly less than those for fibroblasts. The obtained result is in accordance to literature data that states the lower stiffness for cancerous cells compared to the stiffness of the cells of normal tissues [6]. For both analyzed cancerous cell types, the parameters were close to each other, but the significant shift ($p<0.05$) of their values to the lower value interval for HEp-2c cells was fixed (Fig. 2).

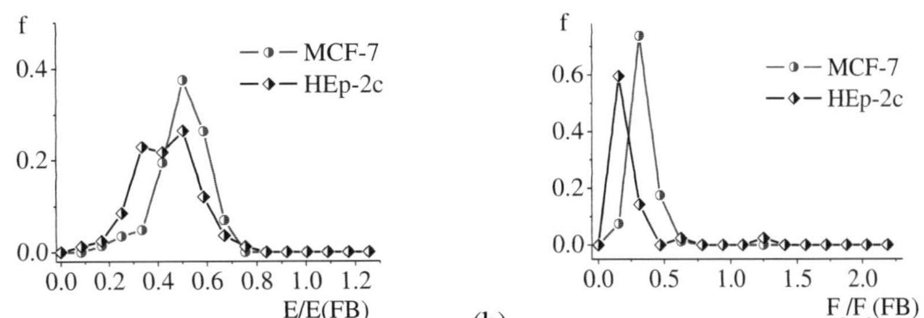

(a) (b)

Figure 2. Frequency polygons of (a) relative elastic modulus and (b) relative adhesive force.

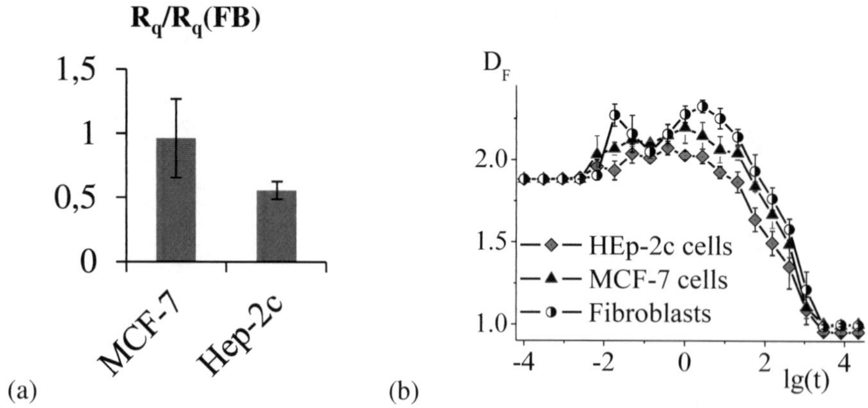

Figure 3. Spatial distribution of the cell mechanical properties. (a) Roughness (R_q) of lateral forces measured over microscale area of the cell surface. (b) Dependence of the fractal dimension (D_F) of microscale lateral force maps for the cells on Z-scale factor (t).

Table 2. Fractal dimension of the lateral force maps for different cell types at two Z-scale coefficients. Data are represented as the limits of 95% CI. n=6-9.

Cell type	t = 0.0183	t = 2.7183
Fibroblasts	2.27±0.07	2.32±0.04
MCF-7 cells	2.07±0.04	2.15±0.08
HEp-2c cells	1.93±0.06	2.02±0.05

The parameters related to the distribution of the mechanical properties of the cell surface were also different for the different cell types. The roughness of lateral force maps for HEp-2c cells was significantly less in comparison with those of both fibroblasts and MCF-7 cells (Fig. 3a).

The parameter $D_F(t)$ for all studied cell types differs both quantitatively and qualitatively (Fig. 3b, Table 2). The fractal dimension of the lateral force maps for HEp-2c cell surface is lower at the coefficients t corresponding to the maxima of dependence $D_F(t)$ for fibroblasts (Table 2) and the shape of the curve $D_F(t)$ is smoother than one for other cell types and has one maximum (Fig. 3b). These facts support more homogeneous distribution of the mechanical properties of the studied surface areas for HEp-2c cells than for MCF-7 cells.

Relative elastic modulus (E/E(FB)) and relative adhesive force ($F_a/F_a(FB)$) are calculated with respect to the mean values of elastic modulus and adhesive force for the samples of fibroblasts.

In contrast to the surface of MCF-7 cells and fibroblasts, there are many microscopic cellular membrane protrusions (microvilii) on the surface of the HEp-2c cells (Fig. 1). The microvilli effectively increase the cell surface area and are useful for such cell absorption and secretion. Adherent cancerous cells as a rule do not have a developed actin cap, but have many short microvilli on their surfaces [7] and the increased presence of microvilli correlates closely with the growth potential and metastatic ability of cancerous cells [8]. On the one side, microvilli can unify the mechanical properties of the cell surface at the microscale, because the cell surface is covered completely with one type of actin structures. On the other side, these superstructures are more flexible than the cellular surface itself, which can change the response of the cell surface with microvilli to mechanical stress. Our experiments were performed in air, thus, the behavior of the microvilli in biological liquids can differ. It is known that microvilli are covered with glycoproteins, which supports the adhesion between cells. However, the general features of mechanical response of the cells surface with microvilli may be expected to be the same as we have revealed in our work.

4. Conclusion

Based on the data obtained with atomic force microscope in contact mode, the physical-mechanical images of the cell surface were obtained for two types of the cancerous epithelial cells in glutaraldehyde-fixed and dried states. They included the averaged values of the parameters characterizing elastic, adhesive, and friction forces and the parameters characterizing the spatial distribution (at the nano- and microscales) of these forces. They are the Young's modulus, adhesive force, roughness, and fractal dimension of lateral force maps and its dependence on Z-scale factor. The study of the mechanical properties of cell surfaces has shown that the response of microscale areas of the HEp-2c cell surface with numerous microvilli to external mechanical forces is less expressed (lower elastic, adhesive, and friction forces) and more homogeneous in comparison with the response of MCF-7 cell surface.

References

[1] Kuznetsova T G, Starodubtseva M N, Yegorenkov N I, Chizhik S A and Zhdanov R I 2007 *Micron* **38** 824
[2] Gladilin E, Gonzalez P and Eils R 2014 *J. Biomech.* http://dx.doi.org/10.1016/j.jbiomech.2014.05.020
[3] Park S 2016 *J. Cancer Prev.* **21** 73
[4] Starodubseva M, Chizhik S, Yegorenkov N, Nikitina I and Drozd E 2010 Study of the mechanical properties of single cells as biocomposites by atomic force microscopy *Microscopy: science. technology. applications and education* vol 1,ed A O Mendez-Vilas and J Diaz (Badajoz. Spain: Formatex Research Center) pp 740-747
[5] Starodubtseva M N, Starodubtsev I E and Starodubtsev E G 2017 *Micron* **96** 96
[6] Lekka M, Pogoda K, Gostek J, Klymenko O, Prauzner-Bechcicki S, Wiltowska-Zuber J, Jaczewska J, Lekki J and Stachura Z 2012 *Micron* **43** 1259
[7] Haghparast S M, Kihara T and Miyake J 2015 *Peer J.* **3** e1131
[8] Ren J, Hamada J, Okada F, Takeichi N, Morikawa K, Hosokawa M and Kobayashi H. 1990 *Jpn. J. Cancer Res.* **81** 920

AFM study of the supramolecular transformation of polyaniline and polyaniline/carbon nanotubes composite upon doping with dodecylbenzenesulfonic acid in the presence of a solvent

I A Lobov*, N A Davletkildeev, D V Sokolov

Omsk Scientific Center SB RAS, Omsk 644040, Russia

*LI__87@mail.ru

Abstract. In this work, we present a new approach for the formation of conducting polyaniline and polyaniline/carbon nanotube composite films with well-developed surface morphology. This approach consists in combined influence of solvent and dopant on the base form of the polymer film. The morphology of the films has been studied by atomic force microscopy. It has been found that polyaniline films possess the fiber structure, and polyaniline/carbon nanotube composite films have the island-like structure.

1. Introduction

Polyaniline (PANI) is one of the most extensively studied conducting polymers due to its unique electrical, optical, chemical properties, stability in air, easy of synthesis, and low cost. PANI is used in a wide range of applications, such as chemical sensors, supercapacitors, display devices, battery electrodes, fuel cells, etc. [1-3]. Physicochemical properties of the PANI may vary within wide ranges depending on synthesis conditions and dopants. The used dopants represent usually strong inorganic and organic acids, such as hydrochloric, sulfuric, camphor sulfonic, methane sulfonic, and dodecylbenzenesulfonic acid (DBSA). The usage of organic acids leads to formation of conductive PANI with increased strength, stability, and conductivity [4-6].

The PANI composites with nanoparticles demonstrate the improved electrical conductivity, tensile strength, and thermostability as compared with pure PANI [1,7]. The unique properties of carbon nanotubes (CNTs) [8] allow using them efficiently as a PANI polymer matrix filler, which leads to significant improvements of the polymer properties [9,10]. There are various methods to obtain PANI/CNTs composites: the introduction of the nanotubes in the polymer matrix by their direct mixing, nanotube surface modification by polymer layer, or in-situ chemical and electrochemical polymerization of aniline in the presence of CNTs [5,11,12]. Last two methods allow obtaining the most stable bond between polymer and nanotube.

Effective surface area plays a very important role in sensitivity of polymer layers, when PANI and composites on its basis are used for sensor devices. Therefore, the surface control of the obtained polymer film is important. Currently, there are a few basic methods to obtain PANI films and composites on its basis with high surface area [7,13,14]. The disadvantages of these methods are limitation of choice of the dopant and poor control of layer thickness needed for sophisticated equipment.

In this work, we propose to form the PANI films and composite of PANI with CNTs (PANI/CNTs composite) in conductive form by influence of a mixture of the solvent and the dopant on the polymer film in the basic form. This technique allows obtaining a well-developed surface morphology. We can produce films of controlled thickness using spin-coating method. The surface morphology of the PANI and PANI/CNTs composite films was examined by atomic force microscopy (AFM). This method allows obtaining 3D image of polymer layers with high resolution.

2. Experimental

2.1. Synthesis of multi-walled CNTs

Multi-walled CNTs were synthesized by chemical vapor deposition resulting from pyrolysis of toluene vapors at 850^{0}C with added ferrocene as a source of catalyst Fe. CNTs were characterized by transmission electron microscopy and AFM [15].

2.2. Synthesis of polyaniline

Solution A: aniline $C_6H_5NH_2$ (2 ml), LiCl (3,2 g), 2-propanol (20 ml), and 3M HCl (21 ml) were mixed and cooled to -15°C. Solution B: ammonium persulfate $(NH_4)_2S_2O_8$ (APS) (1,25 mg), LiCl (1,6 g), and 3M HCl (12 ml) were mixed and cooled to -15°C. LiCl was added to the reaction mixture in order to avoid its congealing. Then, solution B was added slowly dropwise to the solution A with constant stirring in order to initiate the polymerization of the aniline monomer. During polymerization of aniline, the originally colorless mixture turned blue and later became dark green as a result of 1 h reaction. The dark green color indicated formation of the doped emeraldine salt. The product was washed thoroughly with 3M HCl to remove the unreacted monomer and oxidant, and was finally washed with acetone.

2.3. Synthesis of PANI/CNT nanocomposite

PANI/CNTs composite was synthesized by in situ chemical oxidative polymerization of aniline. The CNTs (wt-1% based on aniline monomer concentration) were previously dispersed in the 2-propanol, using sonication for 2 h. Then suspension was dispersed in aniline solution. The reaction mechanism was the same as for PANI.

2.4. Formation of PANI and PANI/CNT nanocomposite developed layers

Emeraldine base (EB) PANI and EB PANI/CNTs composite were dissolved in N-methyl-2-pyrrolidone C_5H_9NO (NMP) and deposited on the Si/SiO_2 substrate via spin coating. To convert EB films into the emeraldine salt (ES) they were doped with dodecylbenzenesulfonic acid $C_{12}H_{25}C_6H_4SO_3H$ (DBSA), mixed with NMP. After that, the film was carefully washed with deionized water and dried in air.

2.5. Measurements

The surface morphology of the PANI and PANI/CNTs composite films was examined by an atomic force microscope MFP-3D SA (Asylum Research) in tapping mode in air. The cantilevers NSG01 (NT-MDT) were used, with resonant frequency 150 kHz, spring constant 5 N/m, and tip curvature radius 10 nm.

3. Morphology studies

Figure 1 shows AFM image of the EB PANI (Fig. 1a) and EB PANI/CNTs composite (Fig. 1b) films. The surface roughness of these films is 3.1 and 4.6 nm for the polymer and composite, respectively. The films structure has a granular appearance. The granule size is 60 ± 7 and 80 ± 12 nm for EB PANI and EB PANI/CNT films, respectively (Fig. 2). The presence of the granules indicates that the polymer molecules are coiled into a ball, which is typical for many macromolecules.

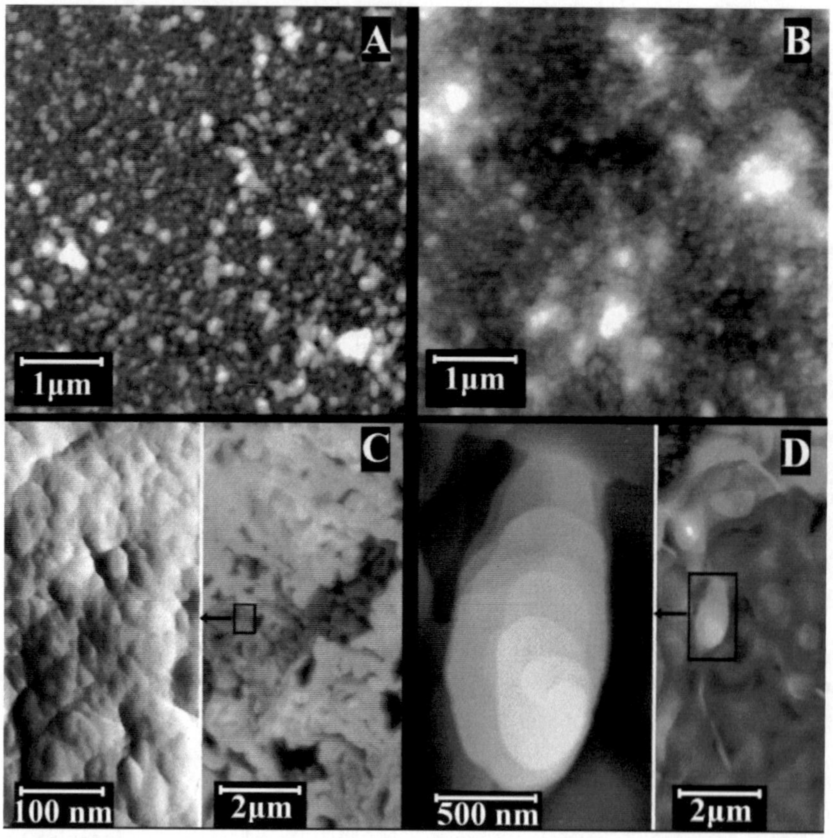

Figure 1. AFM images of: (a) EB PANI, (b) EB PANI/CNTs composite, (c) ES PANI, and (d) ES PANI/CNTs composite films.

After doping EB PANI film, a layer of chaotic interwoven fibers is formed (Fig. 1c). The width of these fibers varies from 200 to 350 nm. The surface roughness of ES PANI film has increased almost by an order of magnitude (22 nm) in comparison with EB PANI film. The observed granular structure of the fibers (Fig. 1c, left part of the image) has an average granule size of about 40 ± 4 nm (Fig. 2). Reducing of the granule size is associated with both an increased packing density of the PANI molecules due to electrostatic force and the transition from hydrophilic to hydrophobic state, which leads to displacing water molecules from the polymer.

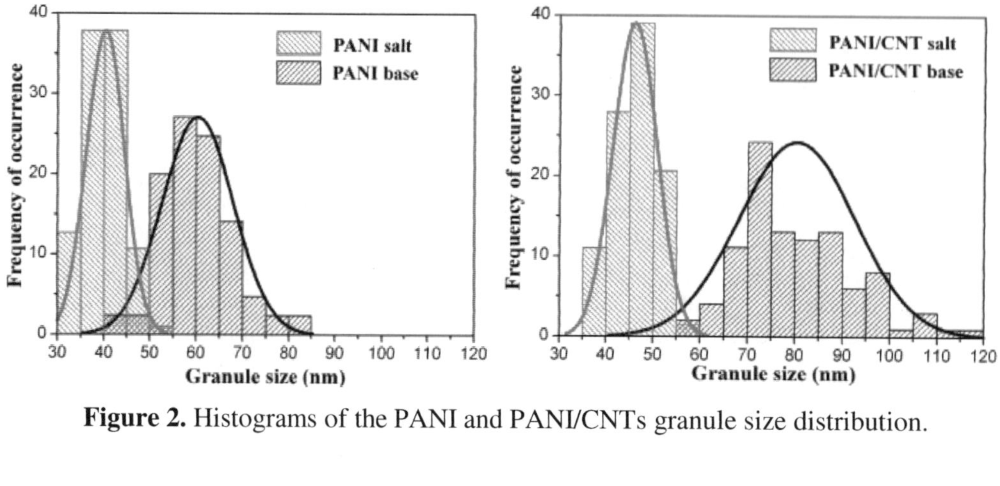

Figure 2. Histograms of the PANI and PANI/CNTs granule size distribution.

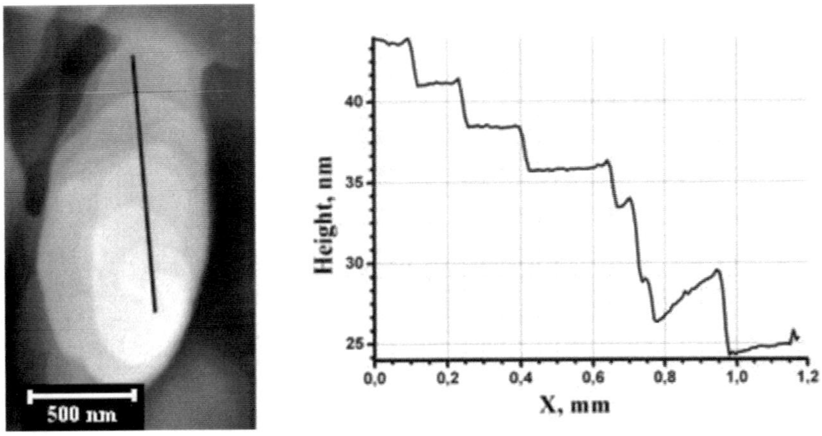

Figure 3. AFM image of island-like structure of PANI/CNT composite and cross section along the line.

After doping EB PANI/CNTs composite film, a developed layer with an island-like structure is formed (Fig. 1d). The islands have a layered structure and consist of superposed circular planes (Fig. 1d, left part of the image). Lack of granular structure indicates that polymer molecules unfold completely. The steps with height from 3 to 45 nm can be observed on the edges of the planes (Fig. 3). A quite large DBSA molecule with a long hydrocarbon tail is used as a dopant. It is interesting, that the height of the steps on the edge of the planes is multiple of length of DBSA molecules. It may indicate the incorporation of the dopant molecules between polymer planes.

It should be noted that the obtained films have areas of the discontinuity of the developed polymer layer (Fig. 4). These areas have a granular appearance (Fig. 4, upper parts of the images). The granule size is 40 ± 4 and 45 ± 4,7 nm for the polymer and composite, respectively (Fig. 2). The surface roughness of such area increases in comparison with EB films (4.2 and 12 nm for PANI and PANI/CNTs composite, respectively). This granular layer is an intermediate layer between the substrate and the developed polymer layer. It occurs due to the strong adhesive interaction between the polymer and the substrate. The layer provides a large contact area and high adhesive force of the developed polymer layer with the substrate.

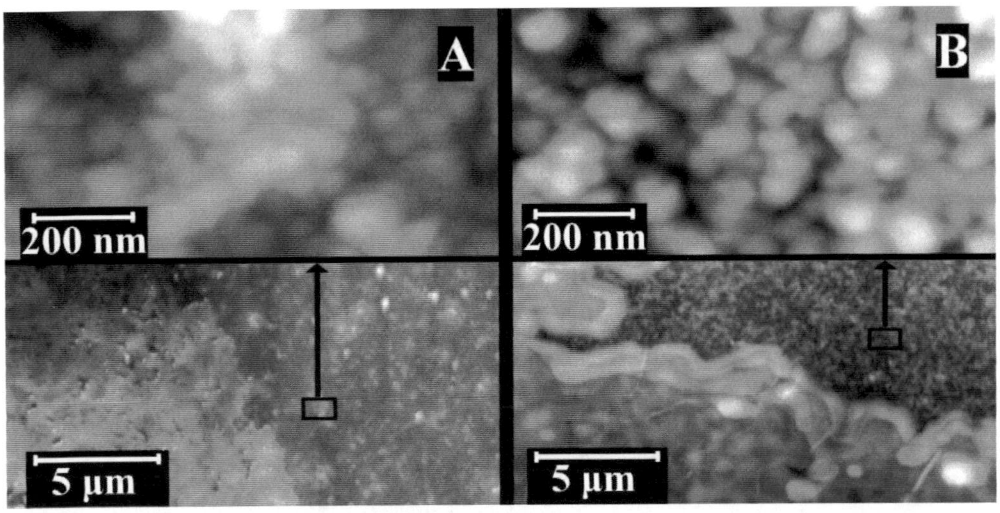

Figure 4. AFM images of ES PANI (a) and ES PANI/CNTs composite (b) films in the areas of the discontinuity of the developed polymer layer.

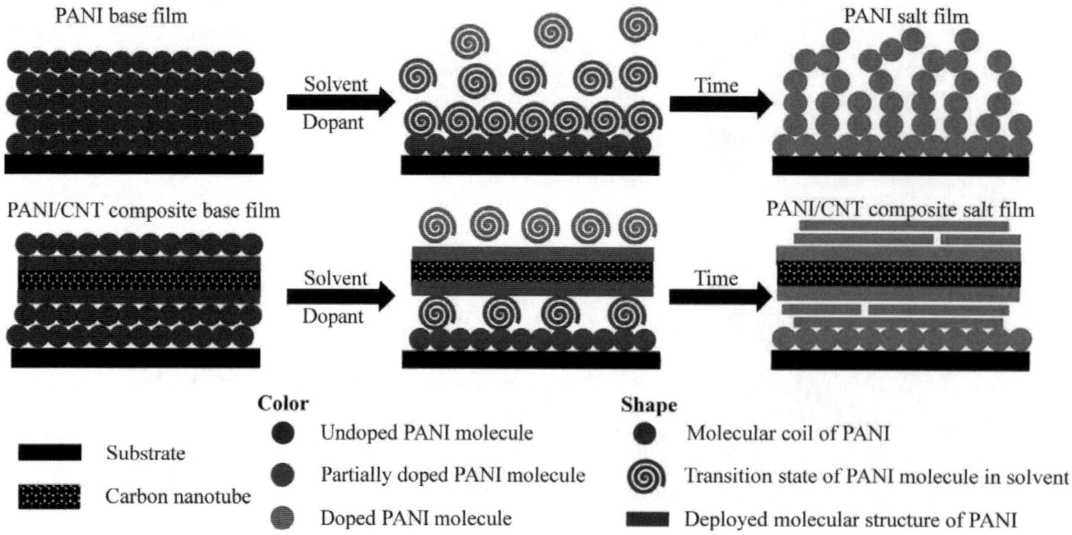

Figure 5. The formation scheme of the developed surface structure. The effect of the solvent on the polymer leads to a change of the molecule shape. The effect of the dopant on the polymer leads to a change of color (from blue to green).

On the basis of the AFM studies, the model of the formation of the developed surface structure can be proposed (Fig. 5). There are processes of dissolution of the polymer surface layer and the doping of polymer chains under the joint impact of solvent (NMP) and dopant (DBSA) on PANI and PANI/CNTs composite films. Under the influence of the solvent, an area with the polymer concentration gradient in the direction of the substrate is formed. The presence of the dopant in the solvent leads to the formation of conductive polymer complexes with DBSA. These complexes are not soluble in NMP. Due to the high concentration of PANI in the subsurface layer of films, active interaction occurs between the polymer molecules, as a result a developed surface morphology of the films is formed. The main contribution to these interactions is made by electrostatic forces. These forces are generated because of the dipoles in the polymer chains appeared by implantation of an acid molecule in the polymer chain. The adhesive force of the polymer molecules to the substrate is high enough that prevents removing the polymer completely from the substrate. In this regard, the layer adjoining the substrate retains the granular appearance and does not undergo changes in supramolecular structure.

It is believed that the chemical oxidative polymerization of aniline in presence of CNTs leads to the formation of PANI with a flat polymer chain configuration. This is mainly due to the interaction between the aromatic structure of polymer and the graphene plane of CNTs due to π-stacking [16,17]. Thus, the CNTs surface is covered by polymer molecules with a deployed molecular structure [18]. Flat polymer chain structure facilitates the transition of molecular coils into ribbons upon doping. We believe that the growth centers of the islands are the CNTs. A polymer layer covering the nanotubes serves as a template.

4. Conclusion

We propose a new approach for the formation of a developed surface of polyaniline (PANI) and polyaniline/carbon nanotubes PANI/CNTs composite films, doped with dodecylbenzenesulfonic acid (DBSA). Initially non-developed emeraldine base (EB) layers transform in fiber and island-like structure for PANI and PANI/CNTs composite, respectively. The developed surface morphology of the films is achieved through simultaneous processes of dissolution of the PANI in the N-methyl-2-pyrrolidone C_5H_9NO (NMP) and doping with DBSA. Polymer films have areas of the discontinuity of the developed polymer layer with a granular appearance. Granule sizes of PANI fibers and non-developed layers are equal.

Acknowledgments

This work was partially supported by the Russian Foundation for Basic Research (grant 16-08-00763a).

References

[1] Ciric-Marjanovic G 2013 *Synth. Met.* **177** 1-47
[2] Song B, Tuan C-C, Huang X, Li L, Moon K-S and Wong C-P 2016 *Mater. Lett.* **166** 12-5
[3] Ansari M O, Khan M M, Ansari S A, Amal I, Lee J and Cho M H 2014 *Mater. Lett.* **114** 159-62
[4] Babu V J, Vempati S and Ramakrishna S 2013 *Mater. Sci. Appl.* **4** 1-10
[5] Jelmy E J, Ramakrishnan S, Rangarajan M and Kothurkar N K 2013 *Bull. Mater. Sci.* **36** 37-44
[6] Li J, Fang K, Qiu H, Li S and Mao W 2004 *Synth. Met.* **142** 107-11
[7] Patil D S, Pawar S A, Devan R S, Ma Y R, Bae W R, Kim J H and Patil P S 2014 *Mater. Lett.* **117** 248-51
[8] Ruoff R S, Qian D and Liu W K 2003 *C. R. Phys.* **4** 993-1008
[9] Gajendran P and Saraswathi R 2008 *Pure Appl. Chem.* **80**(11) 2377-95
[10] Bal S and Samal S S 2007 *Bull. Mater. Sci.* **30**(4) 379-86
[11] Yun S and Kim J 2007 *Synth. Met.* **157** 523-8
[12] Kulkarni M V and Kale B B 2013 *Sens. Actuators B* **187** 407-12
[13] Ma X, Wang M, Li G, Chen H and Bai R 2006 *Mater. Chem. Phys.* **98** 241-7
[14] Lin Q, Li Y and Yang M 2012 *Sens. Actuators* B **161** 967-72
[15] Davletkildeev N A, Stetsko D V, Bolotov V V, Stenkin Y A, Korusenko P M and Nesov S N 2015 *Mater. Lett.* **161** 534-7
[16] Huang J-E, Li X-H, Xu J-C and Li H-L 2003 *Carbon* **41** 2731-6
[17] Wang R-X, Huang L-F and Tian X-Y 2012 *J. Phys. Chem.* C **116** 13120-6
[18] Wang Q, Yao Q, Chang J and Chen L 2012 *J. Mater. Chem.* **22** 17612-8

Combination of scanning probe microscopy techniques for evaluating the electrical parameters of individual multiwalled carbon nanotubes

D V Sokolov[1,*], N A Davletkildeev[1,2], V V Bolotov[1], I A Lobov[1]

[1]Omsk Scientific Center SB RAS, 644024, Omsk, Russia
[2]F.M. Dostoevsky Omsk State University, 644077, Omsk, Russia

[*]classicsub-zero@mail.ru

Abstract. Using two techniques of scanning probe microscopy, the electrical properties (work function, Fermi level position, free carriers' concentration, electrical resistance, conductivity, and carriers' mobility) of individual multiwalled carbon nanotubes were evaluated.

1. Introduction

Carbon nanotubes (CNTs) with unique electrical properties are considered as promising materials for nanoelectronics [1]. Nitrogen doping is an effective way of modifying the electrical properties of CNTs [2], because the nitrogen atoms generate the defects in CNT walls.

For development of the devices consisting of single nanoscale objects (such as nanotubes, nanowires, and nanoparticles), it is necessary to study their properties. Scanning probe microscopy (SPM) is an effective way to obtain information about the properties of individual CNTs [3]. The combination of several SPM methods allows obtaining a complete set of electrophysical parameters of CNTs.

In this work, we evaluate electrophysical parameters of individual undoped and nitrogen-doped multiwalled CNTs based on electrostatic force microscopy (EFM) and conductive atomic force microscopy (C-AFM). To study the change in the electrical properties of CNTs during gas adsorption, SPM measurements of doped CNTs exposed to ammonia (reducing agent) and nitrogen dioxide (oxidizer) gases were carried out.

2. Material and methods

2.1. Multiwalled carbon nanotubes

Chemical vapor deposition (CVD) method was used for synthesis of CNTs. Undoped and nitrogen-doped multiwalled CNTs were grown on SiO_2/Si substrates from pyrolysis of toluene and acetonitrile, respectively. Fe-particles formed by ferrocene decomposition in the synthesis zone at 850°C served as a catalyst for nanotube growth. According to X-ray photoelectron spectroscopy, the nitrogen content in doped CNTs was 3.7 at.%

The samples for the measurements were CNTs deposited on comb-shaped arrays of gold electrodes. Structures with gold contacts were photolithographically formed on a 130 nm thermally grown SiO_2 layer on the n-Si wafer with electrical resistivity 0.1 Ω·cm. CNTs were deposited on gold electrodes from a suspension after ultrasonic dispersion in dichloromethane for 1 hour.

IOP Conf. Series: Materials Science and Engineering **256** (2017) 012018 doi:10.1088/1757-899X/256/1/012018

2.2. Electrical measurements

Individual CNTs lying on two adjacent gold contacts were revealed using AFM-scanning. I-V curve measurements were performed by C-AFM; mean value and range of the longitudinal electrical conductivity of CNTs were determined. Considering the length and outer diameter of CNTs obtained from AFM-images the specific conductivity of CNTs was calculated.

The Fermi level shift in CNTs was estimated using EFM. The measurements were made in the region between the two gold contacts, where the single nanotubes lay on the SiO_2 surface. SPM measurements were performed by the atomic force microscope MFP-3D SA (Asylum Research) in air and in nitrogen atmosphere with reducing (ammonia) and oxidizer (nitrogen dioxide) gases.

3. Results and discussion

Inset in Figure 1 shows the example of the 3D AFM image of a nanotube with diameter 8 nm located between the two adjacent gold electrodes.

The current–voltage curves of Au-CNT-Au structures at voltages from -1 to +1 V have a linear shape. The slope of I–V curve was used to determine the longitudinal electrical resistance of the CNTs on metallic contacts. Resistance of individual CNT was calculated by method considering the contact resistance between nanotube and gold [4]. Figure 1 shows calculated specific conductivity of individual CNTs using the AFM image data for diameter and length of the CNTs. The cross section of the outer wall of nanotube was taken as the cross-sectional area for specific conductivity.

Based on EFM images (Fig. 2, inset), mean value of work function (WF) in CNTs was measured as shown in [5]. For reference, the mean values of WF in undoped and doped CNTs (Fig. 2) and mean values of Fermi level shift were determined, and the carrier concentration of CNTs was calculated (Table 1). The concentration and conductivity values were used to calculate the carrier mobility. In the Table 1 all the electrical parameters of CNTs are presented.

According to the data in Table 1, the conductivity of N-doped CNTs is more than the one of undoped CNTs. This is due to electrically active nitrogen-containing defects, which increase the carrier concentration in doped CNTs and the concentration of electron scattering centers. Therefore, the carrier mobility in doped CNTs is less than in undoped ones.

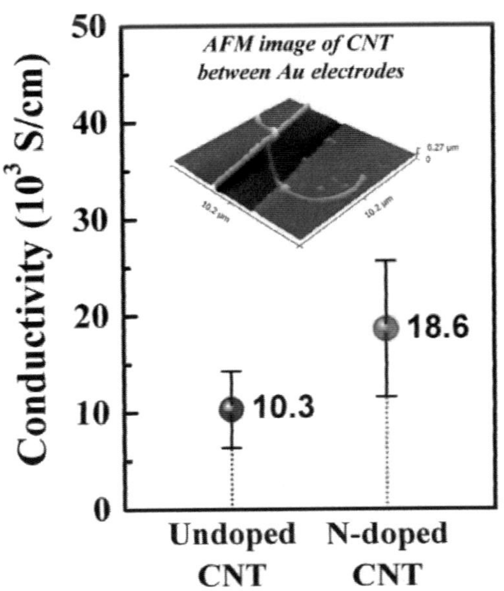

Figure 1. Mean values and range of the longitudinal electrical conductivity of undoped and N-doped CNTs. Inset: 3D AFM image of CNT on gold electrodes.

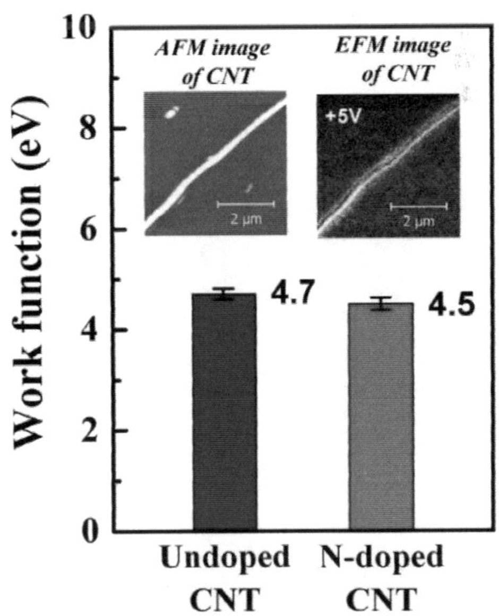

Figure 2. Mean values and range of the work function of undoped and N-doped CNTs. Inset: AFM (left) and EFM (right) image of CNT.

Table 1. Mean value and range of electrical parameters of the individual undoped and N-doped CNTs.

Parameters	The average value and range	
	Undoped CNT	N-doped CNT
Resistance, Ω ·cm	$9.7\cdot10^{-5}\pm3.8\cdot10^{-5}$	$5.4\cdot10^{-4}\pm2.0\cdot10^{-5}$
Conductivity, S/cm	$10.3\cdot10^{3}\pm4.0\cdot10^{3}$	$18.6\cdot10^{3}\pm7.0\cdot10^{3}$
Work function, eV	4.7±0.11	4.5±0.12
Charge carriers concentration, cm^{-3}	$7.9\cdot10^{18}\pm2.8\cdot10^{18}$	$3.2\cdot10^{19}\pm1.1\cdot10^{19}$
Mobility, $cm^2/(V\cdot s)$	$8.16\cdot10^{3}\pm3.2\cdot10^{3}$	$3.6\cdot10^{3}\pm0.3\cdot10^{3}$

The change in the electrophysical parameters of nitrogen-doped CNTs during the adsorption of molecules of the reducing gas (NH_3) and oxidizing gas (NO_2) at concentration of 1000 ppm was studied (Fig. 3) using presented techniques.

When the reducing gas is adsorbed, the Fermi level shift in doped CNTs increases insignificantly (Fig. 3), which may be due to acceptor-like defects in the CNTs walls. In the case of adsorption of the oxidizing gas, the Fermi level shift is substantially reduced (Fig. 3), which indicates the dominance of donor-like defects in the walls of CNTs, in accordance with the type of dopant.

4. Conclusion

The electrical parameters of individual undoped and nitrogen-doped CNTs were obtained using a combination of conductive atomic force microscopy and electrostatic force microscopy techniques. It was shown that the nitrogen, being a donor impurity for CNTs, provided excess concentration of electrons. This leads to an increase of the Fermi level position and a decrease of the work function in doped CNTs.

The electrical conductivity and the Fermi level shift during the adsorption of ammonia and nitrogen dioxide were measured for nitrogen-doped CNTs. The increase and decrease in the average value of the conductivity correlate with increase and decrease of Fermi level position leading to a change of carrier concentration in the outer walls of multiwalled CNTs.

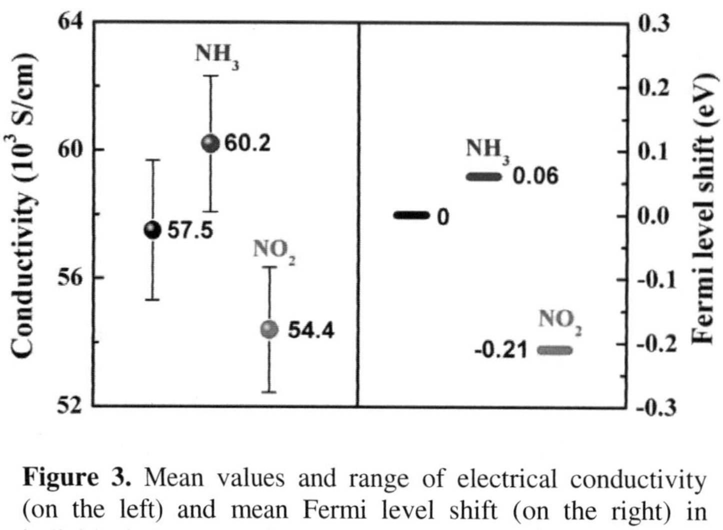

Figure 3. Mean values and range of electrical conductivity (on the left) and mean Fermi level shift (on the right) in individual doped CNTs during adsorption of ammonia and nitrogen dioxide molecules.

IOP Conf. Series: Materials Science and Engineering **256** (2017) 012018 doi:10.1088/1757-899X/256/1/012018

Acknowledgments
This work was partially supported by the Russian Foundation for Basic Research (Grant 15-48-04134 r_siberia_a).

References
[1] Yang N, Chen X, Ren T, Zhang P and Yang D 2015 *Sensor Actuat. B – Chem.* **207** pp 690-715
[2] Ayala P, Arenal R, Rümmeli M, Rubio A and Pichler T 2010 *Carbon* **48** pp 575-86
[3] Mélin T, Zdrojek M and Brunel D 2010 *Scanning Probe Microscopy in Nanoscience and Nanotechnology* ed B Bhushan (Berlin: Springer-Verlag Berlin Heidelberg) ch.4 pp 89-128
[4] An L and Friedrich C R 2012 *Nuclear Instruments and Methods in Physics Research B.* **272** pp 169-72
[5] Davletkildeev N A, Stetsko D V, Bolotov V V, Stenkin Y A, Korusenko P M and Nesov S N 2015 *Materials Letters* **161** pp 534-7

IOP Conf. Series: Materials Science and Engineering **256** (2017) 012019 doi:10.1088/1757-899X/256/1/012019

Scanning probe microscopy in mineralogical studies: about origin of the observed roughness of natural silica-rich glasses

Ye A Golubev* and S I Isaenko

Institute of Geology of Komi SC of Uralian Branch of RAS, Syktyvkar, Russia

*golubev@geo.komisc.ru

Abstract. We have studied different mineralogical objects: natural glasses of impact (tektites, impactites) and volcanic (obsidians) origin, using atomic force microscopy, X-ray microanalysis, infrared and Raman spectroscopy. The spectroscopy showed the difference in the structure and chemical composition of the glasses of different origin. The analysis of the dependence of nanoscale heterogeneity of the glasses, revealed by the atomic force microscopy, on their structural and chemical features was carried out.

1. Introduction

The study of mineral surfaces by the scanning probe microscopy (SPM) methods began about twenty years ago. This is rather late as compared to the application of these methods in other natural science disciplines – physics, chemistry, biology, etc. Natural glasses were one of the first mineralogical objects to study by SPM.

Glasses are melts that are solidified near the glass transition temperature, when the cooling rate exceeds the maximum cooling rate under the given conditions for crystallization [1]. The main geological processes of glass formation are:

(i) explosive and impact processes resulting in the melting of rocks under the influence of ultrahigh pressures (35–90 GPa and above) and high temperatures (up to 3000 °C) as well as formation of *impactites*;
(ii) ablation and transport of the melted substance over significant distances after explosive and impact processes (*tektites* are thus formed);
(iii) melting of rocks at the contact with magma resulting in *buchites*;
(iv) hardening of igneous melts resulting in *obsidians*.

Structural differences due to PT-conditions of cooling, concentration, and composition of impurities, and water content, contribute to the appearance of nanoscale structural peculiarities of the glasses. Although the main SPM studies were carried out for artificial glasses [2-6], the most common natural glasses were also studied by these methods [7-9]. AFM observations of the cleavage of the glasses show a surface of shallow hillocks, which are usually as large as several tens of nanometers wide and several nanometers high. The surface roughness data are actively used to evaluate the processes of corrosion, fracture, and polishing in glasses. However, the nature of this roughness remains unclear. Many authors believe that the presence of roughness indicates that the glasses, including natural ones, are heterogeneous at the nanometer scale [4]. This heterogeneity can be caused both by density fluctuations in pure silicon dioxide, and by composition fluctuations in multicomponent glasses [10]. It has been shown in [11] that the size of the hillocks is much larger in those glasses, in which nuclei of nanocrystallites are present. The possibility of interpreting the cleavage surface of the glasses from the point of view of the discovery of their structural and chemical inhomogeneities, allows speaking about the globular or pseudoglobular structure of the glasses; since for many mineralogical samples, the objects, microscopically observed on the surface, are conditioned by supramolecular structural features [12, 13]. Spectroscopic data allow determining the structure and chemical bonds in natural glasses, distinguishing volcanogenic glasses from impact glasses [14]. The purpose of this study is an attempt to connect structural and chemical data on natural glasses to the results of AFM observations of their surface.

2. Objects and methods

The objects of our study were natural impact glasses: *irgizite* from Zhamanshin Crater (Kazakhstan) and *Libyan Desert glass* (Egypt); volcanogenic *obsidians* (from Kamchatka Peninsula, Armenia, and Italy); and *moldavite* related to tektites (Czech Republic).

The chemical and local structure was determined by the methods of infrared (IR) and Raman spectroscopy, the local chemical composition was estimated by X-ray energy-dispersive spectral analysis. Nanoscale structural features were detected by the atomic force microscopy (AFM).

The surface morphology of the samples has been characterized by AFM measurement in tapping mode using Ntegra Prima microscope (NT-MDT SI, Russia) with super sharp silicon cantilevers of model SSS-NCH (Nanosensors). The resonant frequency of cantilevers is about 330 kHz, the radius at the end is 3–5 nm and the stiffness constant is about 35 N/m. The images have been recorded at a scan frequency between 0.8 and 1 Hz for a resolution of 512×512 pixels. To reduce the effect of static electricity on the images, we increased humidity in the room (relative air humidity of about 85%) and grounded the samples. For each sample, we performed at least 5 AFM images with 3–5 different scan areas.

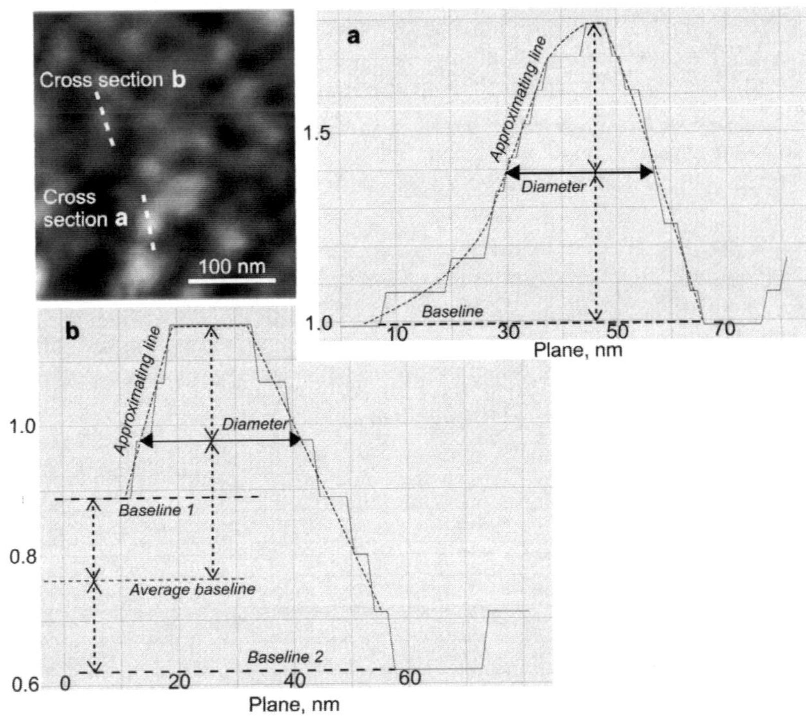

Figure 1. Scheme for measuring of the hillock diameter.

Method for measuring the size of the objects on the surface. In the majority of the works, the surface roughness is used to quantify the objects on AFM images, which is estimated from a standard histogram. In this case, we receive the average height of the hillocks. However, the glasses have a complex surface relief. It can be characterized by roughness of several orders in the image plane: micron-size waves; submicron-size wave; islands of 200–300 nm in diameter, covered by hillocks with diameter of the tens of nanometers. Therefore, the measured roughness always possesses several levels on natural glasses. Moreover, the comparative analysis is complicated by the influence of experimental parameters, such as tip radius, tip-sample force, scan size and pixel resolution. In our work we performed measurements of visible linear sizes (diameter) of the contacting hillocks along the surface profiles (Fig. 1). Due to the complex surface relief, the bases of the hillocks are positioned, as a rule, at the different heights. Therefore, the measurement of diameters was conducted at a half of the hillocks height. The height was measured from the average baseline between baselines of hillock to peak. Such an approach allows decreasing somehow the object broadening effect on the surface by a probe. We measured sizes of 95–120 hillocks for various samples.

The cleaved surfaces of the samples were studied with a Tescan Vega LMH scanning electron microscope with W heated cathode. X-ray energy-dispersive spectrometry (EDS) was used to investigate the chemical composition with AZTEC software (Oxford Instruments) processing.

Raman spectroscopy was carried out with a LabRam HR800 instrument (Horiba, JobinYvon) at room temperature. The system was equipped with an Olympus BX41 optical microscope and a Si-based CCD detector. The spectra were recorded in the 100–4000 cm^{-1} range using a spectrometer grating of 600 g/mm, with a confocal hole of 300 μm and a slit of 100 μm. As an exciting radiation, an external Ar^+ laser (514.5 nm; 1.2 mW) was used. After background correction, individual lines were deconvolved using a curve-fitting procedure from the software provided by LabSpec 5.36.

3. Results

The spectroscopic and X-ray spectral studies allowed establishing the composition of the matrix of glasses (Table 1). It can be seen that SiO_2 content increases in sequence: volcanic glasses – tektite – impactites. On the AFM images the microscopic heterogeneity of the glasses is shown as practically isometric rounded hillocks (Fig. 2), which sizes in the image plane are the tens of nanometers. The distribution of the hillocks is narrow in size, having a lognormal-like appearance (Fig. 3). The smallest average hillock diameters are as follows: for Libyan glass – 21 nm, for irgizite – 26 nm; for moldavite – 34 nm. The average hillock diameter for obsidians varies within 55–70 nm. The hillock nature is traditionally associated with the nanocrystalline structural features of glasses [9,11]; however, published images of high-resolution transmission electron microscopy [15,16] show that atomic lattices in natural glasses are disordered, and it is more correctly to explain their appearance by density fluctuations or by chemical heterogeneity. The results of X-ray spectral analysis of the glasses are given in the Table 1. It can be seen that there is a direct correlation between the hillock size and the content of impurities. The more impurities are, the greater the size of the hillocks is.

Table 1. Composition of the glasses (at. %).

	Libyan glass	Irgizite	Moldavite	Obsidian (Kamchatka)	Obsidian (Italy)	Obsidian (Armenia)
O	66.1	65.5	64.2	61.9	60.9	63.3
Si	33.4	26.5	27.4	24.5	25.5	23.9
Mg	not det.	1.6	1.1	0.9	1.3	0.1
Al	0.3	4.0	3.9	0.8	0.6	6.6
Na	not det.	0.8	0.3	5.8	7.7	2.5
K	–//–	0.8	1.5	0.2	0.2	2.0
Ca	–//–	0.7	1.0	1.9	3.5	0.3
Ti	–//–	0.2	0.1	not det.	not det.	0.1
Fe	0.1	1.4	0.4	0.1	0.1	0.1

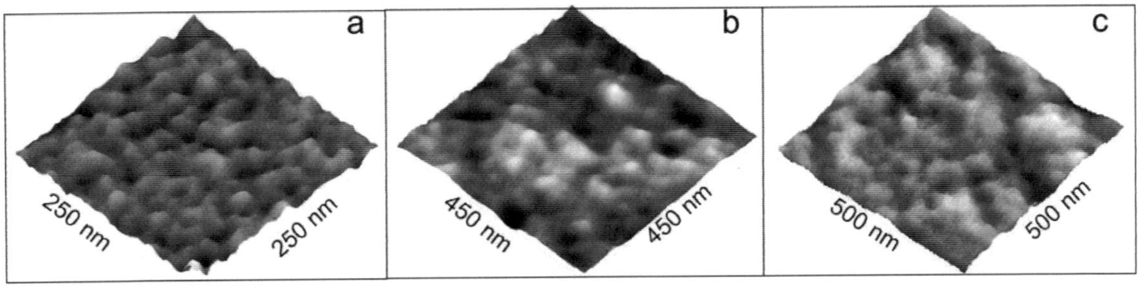

Figure 2. Typical AFM images of the surface of natural glasses: (a) Libyan glass; (b) irgizite; (c) obsidian (Italy).

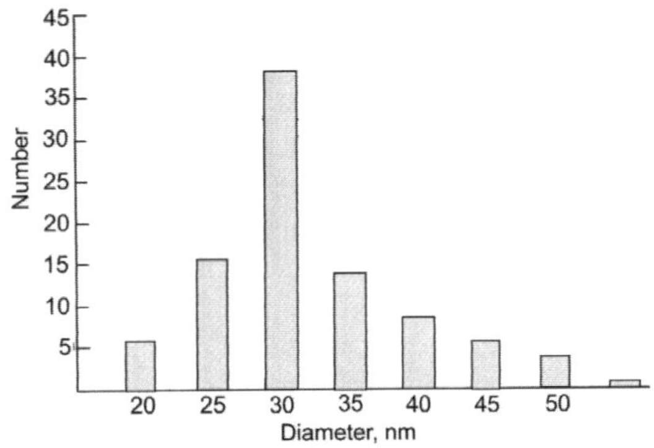

Figure 3. Typical measured distribution of hillocks diameter (for moldavite).

According to IR spectroscopy, three types of vibrations in the SiO_2 molecule were revealed in the glasses. One vibration is associated with crystal structure, in this case with valence vibrations in the quartz molecule. Two others are connected with valence and deformation vibrations in SiO_2 structures.

In our case, a shift of the peak of the valence vibrations of crystalline SiO_2 (900–1100 cm^{-1}) toward smaller angles is observed: in Libyan glass (1102 cm^{-1}), irgizite (1088 cm^{-1}), moldavite (1085 cm^{-1}), and obsidians (1039–1042 cm^{-1}). The absorption bands in this region are typical for all silicate glasses and can be attributed to the valence vibrations of silicon and oxygen atoms in the bridge bonds of the silicon-oxygen tetrahedron. In the spectra of impactites, the absorption band is bifurcated and broadened, which is possibly due to the isomorphous substitution of silicon atoms in the tetrahedron for aluminum atoms. For example, it is known that at the partial replacement of Si^{4+} by Al^{3+}, the valence absorption bands shift to low frequencies. The tetrahedral layer is filled by the principle of the closest packing, so replacing of Si (with ionic radius 0.39 Å) by Al (0.57 Å) or Na (0.92 Å) increases the dimensions of the tetrahedra. Therefore, the shift of this maximum results most likely from the complexity of tektite and volcanic glass. We observe dependence of the diameter of the hillocks on the surface of the glasses on the position of the given maximum in the IR spectrum (Fig. 4). Taking into account the chemical nature of this shift, together with the results of elemental analysis, it is possible to assume the chemical nature of nanoscale heterogeneity of glasses. It can be associated with such a feature of the formation of glasses from melts, as the accumulation of a glass-forming agent (SiO_2) in domains surrounded by a melt with a higher content of cation-modifiers, which are usually elements of the first and second groups [14]. In our samples, there are Na, Ca, Al, and Mg impurities (table), and their content increases from impact glasses to tektites and volcanic glasses. Cations of modifiers are located in free cavities of the structural lattice, compensating excessive negative charge of the complex anion. The strength of the modifier-oxygen bond is much lower than the strength of the glass-to-oxygen bond, so the modifiers do not form strong coordination groups, and when the glass breaks, the bonds break along the clusters of modifier elements.

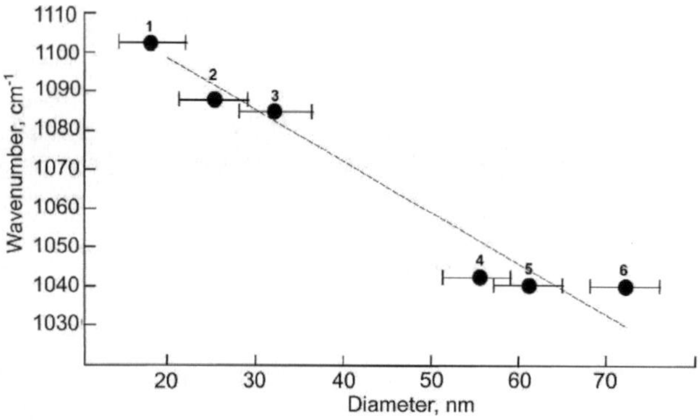

Figure 4. Dependence of the displacement of the ν (SiO_2) band about 1100 cm^{-1} on hillocks diameter. 1 – Libyan Desert glass; 2 – irgizite; 3 – moldavite; 4 – obsidian from Kamchatka; 5 – obsidian from Italy; 6 – obsidian from Armenia.

On the Raman spectra, the structural features of glasses are shown, first of all, in the high-frequency region 800–1200 cm^{-1} [14]. The band with a maximum in the region of 800–850 cm^{-1} is related to the valence vibrations of isolated SiO_4 tetrahedra, in which all the oxygen atoms are non-bridging (structural units Q^0). The bands with a maximum in the regions 900 cm^{-1} and 950–980 cm^{-1} are conditioned by vibrations of terminal groups in SiO_4 tetrahedra, with three and two non-bridging oxygen atoms (structural units Q^1 and Q^2), respectively. The band with a maximum in the range 1050–1200 cm^{-1} is associated with valence vibrations of the non-bridging bonds in SiO_4 tetrahedra with one non-bridging oxygen atom (structural units Q^3). In the low-frequency region (400–700 cm^{-1}) there is an intense band, the maximum of which shifts toward higher frequencies with increasing degree of polymerization of the glass.

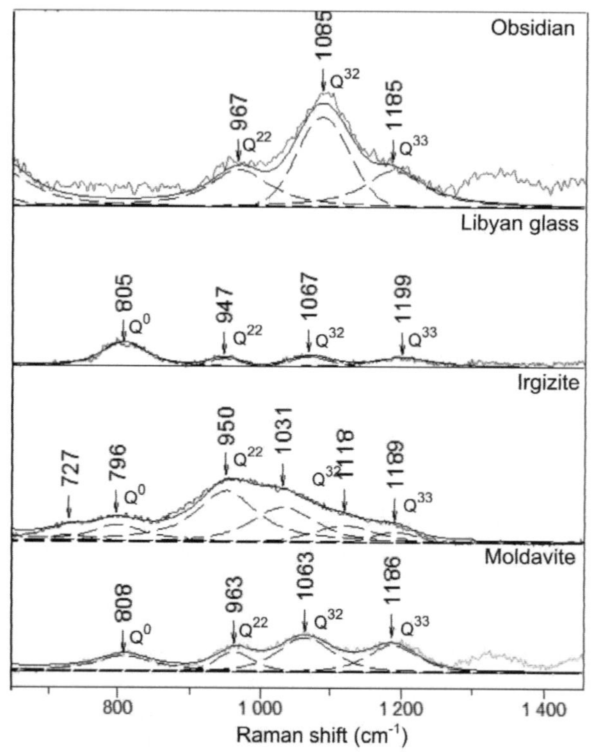

Figure 5. Raman spectra of glasses.

Our samples show mixed distributions of Q^n, and the obsidian spectrum shows the most 3D-ordered structure (Fig. 5). In their spectra, the largest statistical weight of the structural units Q^{32} and Q^{33} is recorded. In the spectrum of moldavite, as in obsidians, the band, corresponding to the structural unit Q^{32}, is predominant, but this band is not yet so clearly distinguished against the background of the remaining bands. In addition, unlike the obsidians, the bands appear, which correspond to the structural units Q^0, which testifies to the significant number of SiO_4-chain breaks and a relatively more disordered structure. In irgizite, the band of the structural unit Q^{22} becomes the predominant one, while the essential role of the band Q^0 remains, which testifies to the relatively smaller ordering of irgizite in comparison with the moldavite. In Libyan glass, the band of the structural unit Q^0 becomes the dominant band, which testifies to the lowest orderliness of this glass in comparison with the others, in particular, to the shortest chains of tetrahedra.

4. Conclusion

According to the results of elemental X-ray spectral analysis and IR spectroscopy, the impact glasses (impactites and tektites) have more homogeneous silica matrix with fewer impurities than volcanic glasses. The dependence of the saturation degree of the natural glasses structure with cation-modifiers (Al, Na, Ca, Mg) on the dimensions of surface inhomogeneities is revealed. It allows relating IR spectroscopic and X-ray spectroscopic data with nanostructural glass features. There is the correlation between the degree of ordering of the structure (according to the Raman spectroscopy data) and content of the impurities in silica glasses. This fact allows suggesting chemical nature of the glass nanoheterogeneity obtained by AFM. The presence of impurities, partially framing the nanosized regions of the glass-forming matrix, promotes the cleavage along the impurity-containing regions and the formation of a rough cleavage surface.

Acknowledgments

This work was supported by the RSF (project No. 17-17-01080). The authors thank G.N. Lysyuk for providing the samples of impactites and moldavite, and V.A. Radaev for the AFM measurements.

References

[1] Henderson G S 2005 The structure of the silicate melts: glass perspective *Canadian Mineralogist* **43** 1921-58

[1] Wünsche C, Rädlein E and Frischat G H 1997 Glass fracture surfaces seen with an atomic force microscope *Fresenius J. Anal. Chem.* **358** 349-51

[2] Raberg W and Wandelt K 1998 Atomically resolved AFM studies of an amorphous barium silicate surface *Appl. Phys.* A **66** S1143-6

[3] Hervé A and Daniel A 2000 Ten years of atomic force microscopy in glass research *Ceramics-Silikáty* **44** 121-8

[4] Poggemann J-F, Goß A, Heide G, Rädlein E and Frischat G H 2001 Direct view of the structure of a silica glass surface *J. of Non-Crystal Solids* **281** 221-6

[5] Dalmas D, Lelarge A and Vandembroucq D 2007 Quantitative AFM analysis of phase separated borosilicate glass surfaces *J. of Non-Crystal Solids* **353** 4672-80

[6] Radlein E and Frischat G H 1997 Atomic force microscopy as a tool to correlate nanostructure to properties of glasses *J. of Non-Crystal Solids* **222** 69-82

[7] Heide G, Müller B, Kloess G, Moseler D and Frischat G H 2003 Structural classification of natural non-crystalline silicates *J. of Non-Crystal Solids* **323** 68-71

[8] Frischat G H, Poggemann J-F and Heide G 2004 Nanostructure and atomic structure of the glass seen by atomic force microscopy *J. of Non-Crystal Solids* **345** 197-202

[9] Gaskell P H 1997 Structure and properties of glasses - how far do we need to go? *J. of Non-Crystal Solids* **222** 1-12

[10] Marliere C, Prades S, Celarie F, Dalmas D, Bonamy D, Guillot C and Bouchaud E 2003 Crack fronts and damage in the glass at the nanometer scale *J. of Phys.: Cond. Matter* **15** S2377-86

[11] Golubev Ye A 2003 Scanning probe microscopy in researches of micro- and nanostructure in noncristalline geomaterials *Microscopy and Microanalysis* **9** 304-5
[12] Golubev Ye A 2005 Supermolecular nanostructurization in natural colloids: scanning probe microscopy data *J. of Crystal Growth* **275** e2357-60
[13] Anfilogov N N, Bykov N N and Osipov V N 2005 Silicate melts (Moscow: Nauka) (in Russian)
[14] Gornostaeva T A, Mokhov A V, Kartashov P M and Bogatikov O A 2016 Condensate glasses from the Zhamanshin Crater. I. Irghizites *Petrology* **24** 1-20
[15] Gornostaeva T A, Mokhov A V, Kartashov P M and Bogatikov O A 2017 Condensate glasses from the Zhamanshin Crater. II. Zhamanshinites *Petrology* **25** 3-25

Phase transitions in copper oxide thin films under proton irradiation

M M Dementyeva[1,*], K E Prikhodko[1,2], B A Gurovich[1], Z V Bukina[1], D A Komarov[1] and L V Kutuzov[1]

[1] National Research Centre «Kurchatov Institute», Moscow, 123182, Russia
[2] National Research Nuclear University (MEPhI), Moscow, 115409, Russia

[*]dementyeva_mm@nrcki.ru

Abstract. The variation in the atomic composition of thin CuO films under irradiation by proton beams with different doses was experimentally studied by X-ray photoelectron spectroscopy (XPS), electron energy loss spectroscopy (EELS), and high resolution transmission electron microscopy (HRTEM). The chemical composition and corresponding changes of crystal phases were found after proton irradiation at various doses according to the XPS, EELS, and HRTEM data. It was demonstrated that the selective removal of oxygen atoms was accompanied by a phase transition to Cu_2O during the gradual recovery of the virgin CuO.

1. Introduction

Physical basis underlying the process of selective removal of atoms (SRA) under the low energy proton irradiation of thin films was demonstrated experimentally in [1]. This SRA radiation technology allowed producing various functional nanoelements based on the thin-film materials with the possibility to control the composition and properties in different layers of multilayer structures, as well as recover materials from their chemical compounds without surface sputtering. The selective removal of atoms was carried out at low ions energies in the range of (0.1-4) keV with proton irradiation, which distinguished this process from the effects of selective sputtering of atoms and ion-induced mixing of atoms at the interface between the two phases [2].

In this work, CuO reduction was carried out under proton irradiation. The purpose of this work was to study the evolution of the structural-phase state of the films during low energy proton irradiation. CuO thin films were good model objects to demonstrate the SRA process, because the copper oxides easily underwent the hydrogen irradiation [1]. The SRA process of CuO thin films under proton-reduced irradiation was studied in this work using several techniques: X-ray photoelectron spectroscopy (XPS), electron energy loss spectroscopy (EELS), and high resolution transmission electron microscopy (HRTEM).

2. Materials and experimental methods

Initial samples containing copper and oxygen (CuO) were prepared by cathode sputtering of copper target on a single crystal silicon substrate coated with a 0,15 μm layer of amorphous silicon oxide SiO_2 followed by the oxidation in air at 500°C for 1 h. The thickness of the initial films was about 10 nm.

IOP Conf. Series: Materials Science and Engineering **256** (2017) 012020 doi:10.1088/1757-899X/256/1/012020

CuO thin films were irradiated by ion beams extracted from high-frequency discharge plasma. Proton beam irradiation was performed with particle energies about 1 keV in a dose range $(0.03\text{-}1.4)\cdot 10^{18}(H^+/cm^2)$.

Phase and chemical composition of initial sample CuO were studied using "Titan 80-300ST" transmission electron microscope, operated at 200 kV, equipped with a GIF-2001 energy loss spectrometer. The EEL spectra were obtained with an energy dispersion of 0.5 eV/ch; convergence angle α was 10 mrad, and collection angle β was 14.85 mrad. Cross sections of the samples $CuO/SiO_2/Si$ were prepared by FIB "Helios Nanolab 650".

Quantitative EELS analysis was carried out with equation (1):

$$\frac{N_A}{N_B} = \frac{I_A(\beta,\Delta)}{I_B(\beta,\Delta)} \cdot \frac{\sigma_B(\beta,\Delta)}{\sigma_A(\beta,\Delta)}, \qquad (1)$$

where I_A and I_B are integrated intensities of the EELS peaks after background subtracting, and σ_A and σ_B are ionization cross-sections [3].

The phase composition in the initial and irradiated samples was obtained from the corresponding HRTEM image using Fourier transform diffraction pattern technique similar to the one used in [4].

The identification of different chemical states of copper Cu^{2+}, Cu^{+}, and Cu^{0} in thin films at different ion irradiation doses was performed using XPS taking into account the difference in peaks shape and corresponding chemical shift of energy spectra position. XPS analysis was performed on "Quantera Scanning XPS Microprobe System", using monochromatic Al K-α X-rays. The spectra were obtained with 112 eV pass energy and 0.2 eV energy step size. To eliminate energy shift due to possible surface charge, the spectra were referenced to adventitious C_{1s} peak at 284.8 eV.

3. Results and discussion

The standard copper oxidation method [5] in air for 1 hour at 500^0C allowed obtaining the 10-nm-thick copper oxide layer as determined by HRTEM. Transmission electron microscopy data showed that initial films had polycrystalline structure with different grain orientations. Analysis of diffraction patterns showed that individual grains of initial material corresponded to a monoclinic CuO phase system (C2/c) with the unit cell parameters: a=0.4662 nm, b=0.3417 nm, c=0.5118 nm, $\alpha=\gamma=90^o$, $\beta=99{,}48^o$ (Fig. 1). Elements depth-profile distribution in initial CuO film obtained by EELS technique in STEM mode is shown in Figure 2. Depth distributions of copper and oxygen relative concentrations in Figure 2 were calculated from EELS data using Eq. (1). Figure 2 proved the uniform distribution of the elements through the depth of the film, both average oxygen and copper content in the film was about 50% at.

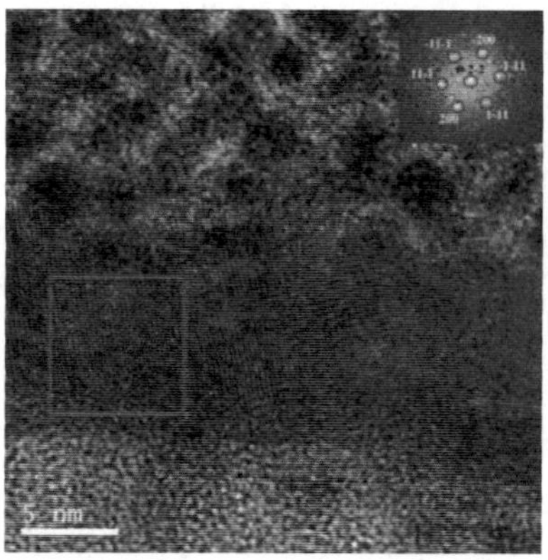

Figure 1. Bright-field HRTEM images of CuO film cross section synthesized on a SiO_2 substrate, in initial state, diffraction (inset).

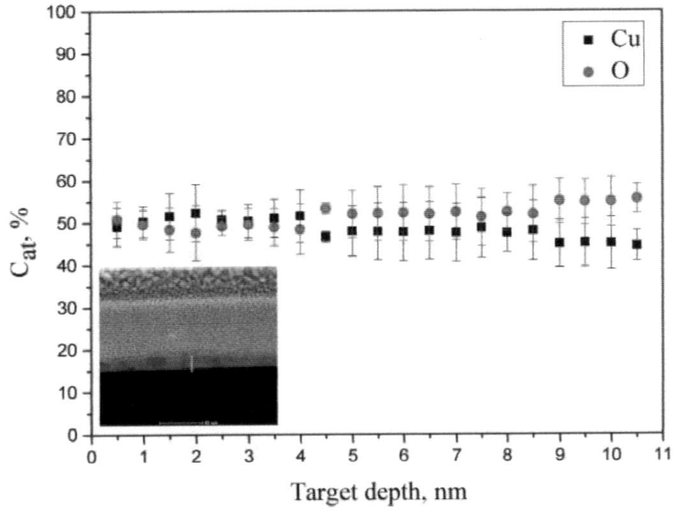

Figure 2. Elements distribution profile on the film depth for one grain (STEM image insert) calculated from EELS analyzing data.

The copper oxide CuO was formed from initial Cu film after thermal oxidation in air. Regardless of the valence state, Cu phase was characterized by $Cu2p_{3/2-1/2}$ doublet. The CuO XPS spectrum is characterized by high-intensity shake-up satellites at 9 eV higher binding energy than the main 2p3/2 and 2p1/2 peaks, which are in addition considerably broader than in the case of bulk CuO and Cu metal [6, 7]. Figure 3a demonstrates a good correspondence for normalized $Cu2p_{3/2-1/2}$ XPS spectra obtained for CuO film in this paper and the one obtained for fully oxidized CuO surface according to [6]. Figure 3b shows XPS spectra of initial CuO film (curve a) compared to the one irradiated up to a dose of $0.7 \cdot 10^{18}$ H^+/cm^2 (curve b).

After proton irradiation of initial CuO film up to $1.4 \cdot 10^{18}$ H^+/cm^2, the phase transition with formation of Cu_2O occurred. Figure 3 (curve a) shows XPS spectra of irradiated CuO film up to a dose of $1.4 \cdot 10^{18}$ H^+/cm^2. Cu2p peak in the initial sample fitted the Cu2+ state [7] with shake up satellites at high binding energies, absent in the spectra of irradiated samples (Fig. 3). Cu2p3/2 peak position for irradiated samples (932.7 eV) as compared to the one for initial state (934.2 eV) also indicated the change of Cu valence state from 2 toward 1 or even 0. Thus, we observed a typical change in the spectrum during SRA process.

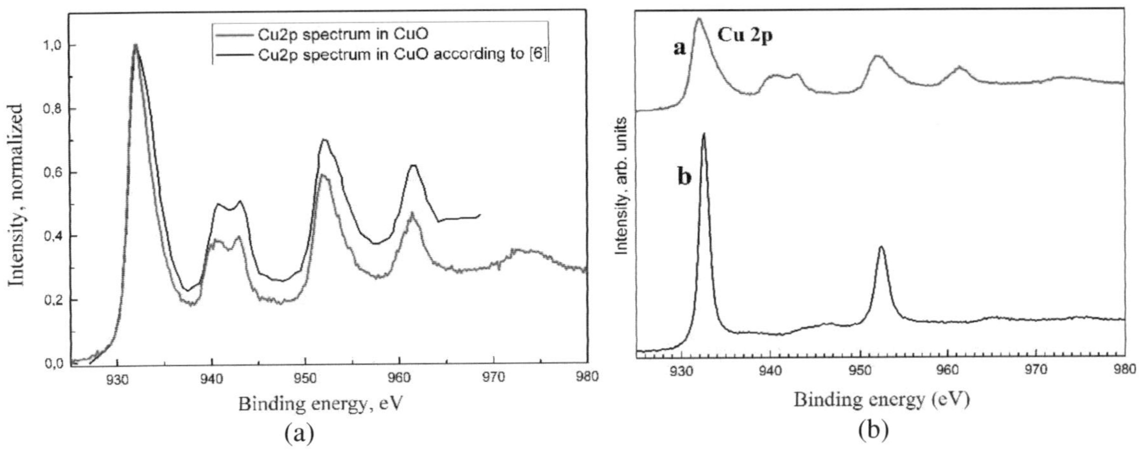

Figure 3. $Cu2p_{3/2-1/2}$ XPS spectra of (a) CuO film obtained in this paper and in [5], normalized for comparison; (b) initial CuO film (curve a) and irradiated up to the dose of $0.7 \cdot 10^{18}$ H^+/cm^2 (curve b).

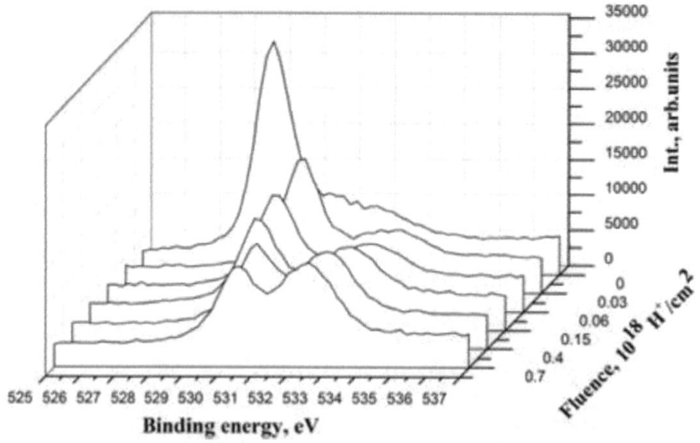

Figure 4. Evolution of O1s signal for Cu-O films: two features account for Cu-O and Si-O states.

It is well known that during the recovery process under ion beam irradiation of the different oxides films the thickness of the films decreases due to the removing of oxygen atoms according to SRA mechanisms [1]. Decreasing of the film thickness during the ion beam irradiation leads to the appearance of the silicon signal from the substrate in XPS spectra. Because of the well-defined chemical Si^{+4} state of silicon in thermal oxide in the substrate, there was easy to separate the part of oxygen signal corresponded to silicon oxidized state to get true oxygen concentration in Cu oxide film during irradiation.

For all the samples, O1s, Cu2p, and Si2p lines were analyzed. Si2p peak position was confined to (103.2-104 eV), indicating the oxide state of silicon. Figure 4 presented the evolution of O1s signal for Cu-O films during irradiation. Nature of the small arm on the part of high binding energy of oxygen O1s peak presented in the initial state is not clear. The shape of O1s peak for irradiated samples showed a complex structure implying at least two components (Si-bonded and Cu-bonded). Evolution of this peak could be characterized by decrease of oxygen component, associated with copper. This effect indicated the implementation of selective removal of oxygen atoms from initial compound of CuO. It was mentioned above that the weak Si2p peak was attributed to SiO_2 substrate. The oxygen total amount was reduced by the value contained in the substrate to estimate the amount of Cu-bonded oxygen.

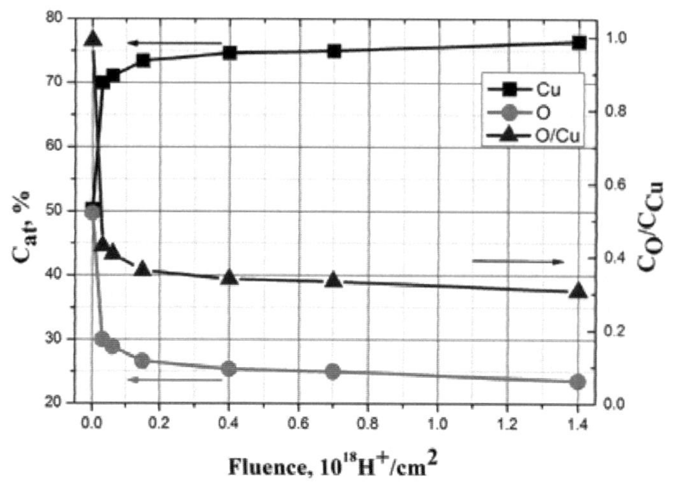

Figure 5. Dose dependencies of Cu and O atomic concentrations and Cu/O atomic ratio for CuO films under 1 keV proton irradiation.

Figure 5 shows O/Cu atomic ratio and corresponding Cu and O atomic concentrations in the film, obtained from XPS spectra analysis. At low irradiation doses, removal of oxygen atoms from the film was accompanied by a phase transition $CuO \rightarrow Cu_2O$. For higher irradiation doses up to $1.4 \cdot 10^{18}$ H^+/cm^2, the O/Cu ratio decreased, followed by the consequent removing of oxygen atoms from the film. Decreasing of the recovery speed with increasing of irradiation dose is a common feature of SRA process kinetics of different materials. It is clear from Figure 5 that complete recovering of CuO film to metal state would require higher doses of ion irradiation.

4. Conclusion

To summarize results obtained in the present work, we can conclude that initial copper films oxidized in air have phase composition CuO. Further irradiation with 1 keV proton beams at different doses leads to phase transition $CuO \rightarrow Cu_2O$, which confirms that the process of selective removal of oxygen atoms takes place under 1 keV proton irradiation.

References

[1] Gurovich B A and Prikhodko K E 2009 *Physics-Uspekhii* **52** 165-78
[2] Taglauer E 1982 *Applications of Surface Science* **13** 80-93
[3] Williams D and Carter A 2009 *Electron microscopy* New York: Springer
[4] Prikhodko K, Gurovich B and Dement'eva M 2016 *IOP Confernce Series: Materials Science and Egineering* **130** 012046-51
[5] Gong Y S, Chiapyng L and Yang C K 1995 *Journal of Applied Physics* **77** 5422-5
[6] Poulston S, Parlett P M, Stone P, Bowker M and Poulston 1996 Surface and interface analysis **24** 811-20
[7] Espinos J P, Morales J, Barranco A, Caballero A, Holgado J P and Gonzalez-Elipe A R 2002 *Journal Physical Chemistry* **106** 6921-9

IOP Conf. Series: Materials Science and Engineering **256** (2017) 012021 doi:10.1088/1757-899X/256/1/012021

Study of the field emission graphene/SiC nanostructures using scanning probe microscopy

I L Jityaev[1,*], A M Svetlichnyi[1], A S Kolomiytsev[1], E Yu Volkov[2], V V Polyakova[1] and O A Ageev[1]

[1]Southern Federal University, Research and Educational Center "Nanotechnology", Taganrog, 347922, Russia
[2]North Caucasus Federal University, Institute of Electric Power Engineering, Electronics and Nano-technologies, Stavropol, 355029, Russia

[*]jityaev.igor@gmail.com

Abstract. We investigated the topology and electrical characteristics of the field emission graphene/SiC nanostructures using scanning probe microscopy. The effect of design of graphene/SiC nanostructures on field emission properties was estimated. The current-voltage characteristics were measured at different rounding-off radii of the emitting top and the interelectrode distances.

1. Introduction
One of the important tasks of nanoelectronics is development of the field emitters with low operating voltages. In this case, an emission cell should satisfy certain geometric requirements and requirements for the cathode material. An emitter must be made of a mechanically strong, heat-resistant material with a low operating function. The electrical and mechanical properties of graphene on SiC make it possible to produce field emission structures with increased resistance to destabilizing factors. Thus, nanoscale graphene/SiC field emitters are promising electron sources [1-6].

At the same time, the development of nanoscale field emission systems requires using special research methods. The scanning probe microscopy (SPM) methods are universal and allow obtaining information about the geometry of nanometer field emitters with subsequent measurement of their electrical characteristics with nanometer locality and high precision [6-10].

The purpose of this work is the application of SPM methods for investigation of field emission cathodes based on graphene films obtained by thermal decomposition of silicon carbide in vacuum.

2. Design and production
It has been found recently that emission structures with nanometer interelectrode distance and the rounding-off radius of the emitter top have a high field gain [11, 12]. The electric field strength reaches values about 10^9 V/m with a potential difference below 10 V. Thus, tip-shaped field emission cathodes with different nanometer rounding-off radius of the emitter top were produced. Silicon carbide 6H-SiC doped with nitrogen with an impurity concentration of 10^{18} cm^{-3} was used as a substrate. Focused gallium ion beams were used for etching of the emission nanostructures [4-6, 13, 14]. This method is characterized by high accuracy of etching, locality, and the ability to fabricate nanometer structures. The liquid etching methods of SiC require high temperatures, longer time, and

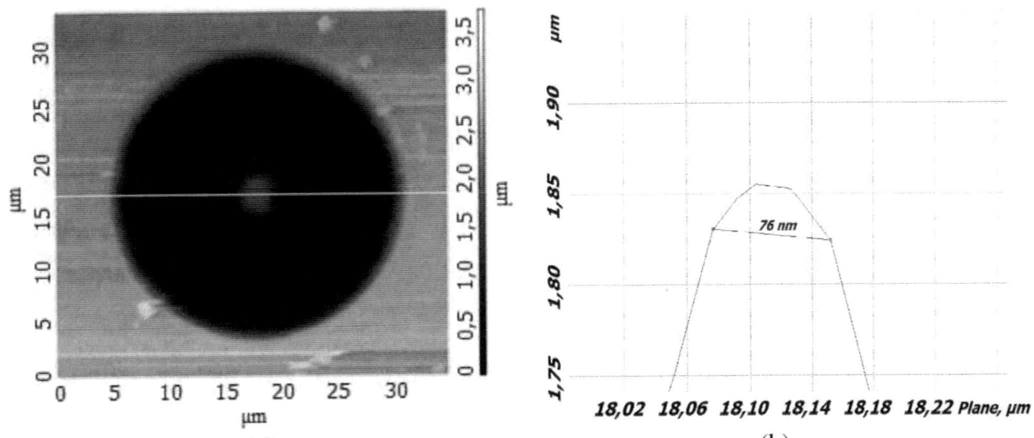

Figure 1. (a) AFM image of the tip-shaped field emission graphene/SiC nanostructure and (b) cross-section of the emitter top along the white line.

the application of more laborious methods of photolithography. Graphene films on the emitter surface were obtained by the thermal decomposition of SiC [1, 4, 5]. This method is manufacturable and allows creating low-defect graphene films on the entire surface of a SiC substrate.

3. Experiment and results

The geometry of field emission cathodes based on graphene/SiC was measured by atomic force microscope Ntegra Vita (NT-MDT Spectrum Instruments). AFM-images of nanostructures showed that a series of emitters with rounding-off radii of the top ranging from 20 to 40 nm were fabricated. The AFM image of the emitter with a 40 nm rounding-off radius of the top is shown in Figure 1.

Field emission from a flat surface was investigated at the first stage to estimate the emission ability of the graphene films on SiC. Current-voltage characteristics of the flat graphene/SiC surface at an interelectrode distance less than 1 nm are shown in Figure 2. The threshold voltage at a cathode-anode distance of less than 1 nm was tenths of a volt. An increase of the interelectrode distance leads to a reduction of the emission current.

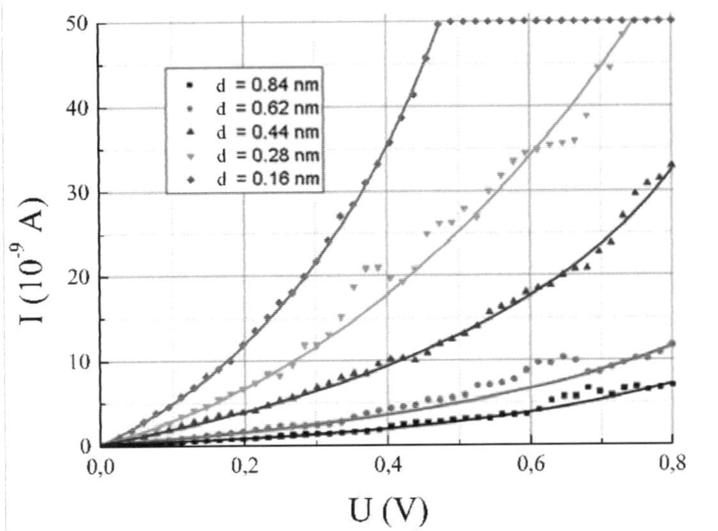

Figure 2. Current-voltage characteristics of the flat graphene/SiC surface.

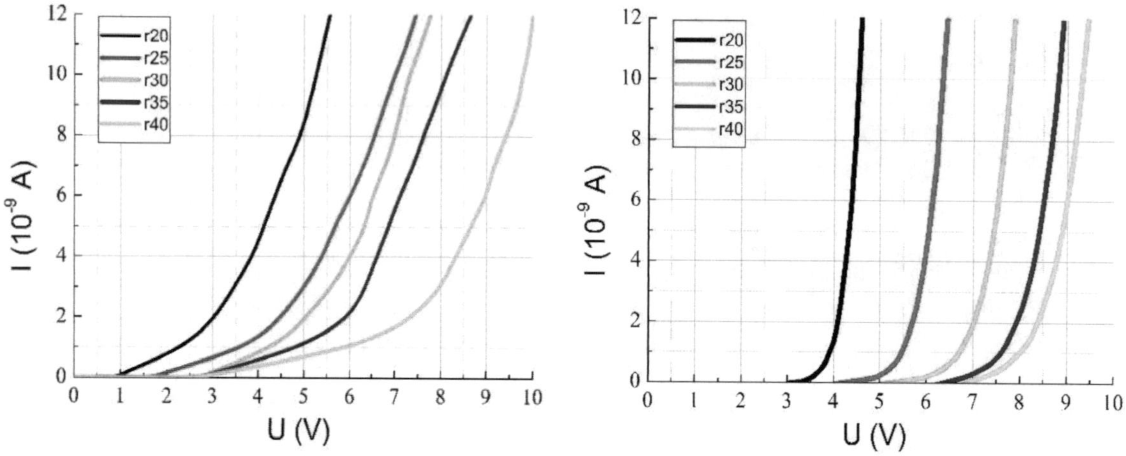

(a) (b)

Figure 3. (a) Experimental and (b) theoretical current-voltage characteristics of the tip-shaped field emission graphene/SiC nanostructure.

Simulation of the tip-shaped field emission nanostructures shows that a decrease of the rounding-off radii of the top leads to an increase in the field gain. The use of SPM for the study of the field emission allows revealing the effect of the nanosized rounding-off radius of the top at a nanometer interelectrode distance on the current-voltage characteristics. The effect of the rounding-off radius of the emitter top on field emission was estimated. The emitters with 5 nm interelectrode distance were investigated experimentally and current-voltage characteristics were simulated theoretically (Fig. 3). It was found that reducing the rounding-off radius of the emitter top from 40 to 20 nm contributed to decrease of the threshold voltage from 2 to 1 V and increase of the emission current.

The current-voltage characteristics in the Fowler-Nordheim coordinates based on the experimental data were plotted (Fig. 4a). The plot slope was used to determine the work function. The dependence of the work function on the rounding-off radius of the emitter top was plotted on the basis of the calculation results taking into account the slope of the F-N plot (Fig. 4b).

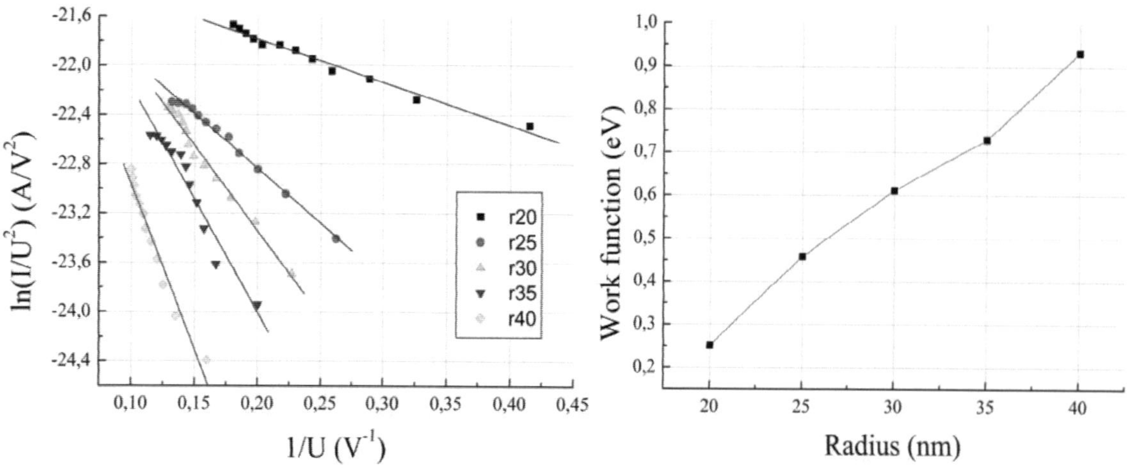

(a) (b)

Figure 4. (a) Experimental Fowler-Nordheim plots and (b) dependence of the work function from a rounding-off radius of the emitter top.

4. Conclusion

It is shown that the application of modern SPM makes it possible to carry out complex studies of nanoscale structures. The emission characteristics at nanometer interelectrode distances were measured by SPM with high localization. The obtained data are in good agreement with the theoretically calculated current-voltage characteristics. The dependence of the work function on the rounding-off radius of the emitter top was obtained on the basis of experimental studies. The work function is 0.25 eV for rounding-off radius of the emitter top of 20 nm. The calculated low values of the work function correspond to the results for emitters based on graphene and graphene-like materials [4, 5, 15]. Our studies of nanoscale field emitters based on graphene films on SiC at nanometer interelectrode distances can be used for the development of energy-efficient vacuum nanoelectronics.

Acknowledgments

This work was supported by The President of the Russian Federation Grants Council (Grant no. MK-6163.2016.8). The equipment of the Center for Collective Use "Nanotechnology" and the Research and Education Center "Nanotechnologies" Southern Federal University was used for this study.

References

[1] Lebedev A A, Kotousova I S, Lavrent'ev A A, Lebedev S P, Makarenko I V, Petrov V N and Titkov A N 2009 *Physics of the Solid State* **51** 829
[2] Wu Ch, Li F, Zhang Y and Guo T 2013 *Vacuum* **94** 48
[3] Cai W, Zeng B, Lui J, Guo J, Li N, Chen L and Chen H 2013 *Applied Surface Science* **284** 113
[4] Konakova R V, Okhrimenko O B, Svetlichnyi A M, Ageev O A, Volkov E Yu, Kolomiytsev A S, Jityaev I L and Spiridonov O B 2015 *Semiconductors* **49** 1242
[5] Konakova R V, Okhrimenko O B, Kolomys A F, Strel'chuk V V, Svetlichnyi A M, Ageev O A, Volkov E Yu, Kolomiitsev A S, Zhityaev I L and Spiridonov O B 2016 *Journal of Superhard Materials* **38** 235
[6] Jityaev I L, Ageev O A, Svetlichnyi A M, Kolomiytsev A S and Spiridonov O B 2016 *Journal of Physics: Conference Series* **741** 012011
[7] Bartolomeo A Di, Scarfato A, Giubileo F, Bobba F, Biasiucci M, Cucolo A M, Santucci S and Passacantando M 2007 *Carbon* **45** 2957
[8] Furkert S A, Wotherspoon A, Cherns D, Fox N A, Fuge G M, Heard P J and Lansley S P 2007 *Applied Physics Letters* **90** 242109
[9] Le Fèbre A J, Abelmann L and Lodder J C 2008 *Journal of Vacuum Science & Technology* B **26** 724
[10] Harniman R L, Fox O J L, Janssen W, Drijkoningen S, Haenen K and May P W 2015 *Carbon* **94** 386
[11] Volkov E, Jityaev I and Kolomiitsev A 2015 *IOP Conf. Series: Materials Science and Engineering* **90** 012031
[12] Jityaev I L, Ageev O A, Svetlichnyi A M and Spiridonov O B 2016 *AIP Conference Proceedings* **1772** 040010
[13] Ageev O A, Kolomiitsev A S and Konoplev B G 2011 *Semiconductors* **45** 89
[14] Avilov V I, Ageev O A, Jityaev I L, Kolomiytsev A S and Smirnov V A 2016 *Proceedings of SPIE* **10224** 102240T-1
[15] Fursey G N, Petrick V I and Novikov D V 2009 *Technical Physics* **54** 1048

IOP Conf. Series: Materials Science and Engineering **256** (2017) 012022 doi:10.1088/1757-899X/256/1/012022

Tribological properties investigation of the thermoplastic elastomers surface with the AFM lateral forces mode

T A Kuznetsova[1,*], T I Zubar[1], V A Lapitskaya[1], K A Sudzilouskaya[1], S A Chizhik[1], A L Didenko[2], V M Svetlichnyi[2], M E Vylegzhanina[2], V V Kudryavtsev[2] and T E Sukhanova[2,3]

[1]A. V. Luikov Heat and Mass Transfer Institute of NAS Belarus, Minsk, Belarus
[2]Institute of Macromolecular Compounds of RAS, Saint-Petersburg, Russia
[3]Federal State Unitary Enterprise "S. V. Lebedev Institute of Synthetic Rubber", Saint-Petersburg, Russia

[*]kuzn06@mail.ru

Abstract. The series of new thermoplastic elastomer films based on copoly(urethane-imide)s (coPUI)s and nanocomposites containing from 1 to 10 wt. % carbon nanofillers of different morphology (single-walled carbon nanotubes, carbon nanofibers, and graphene) as well as WS_2 and WSe_2 nanoparticles, were prepared and investigated by atomic force microscopy in contact mode. The friction coefficient (C_{fr}) on the films surfaces under conditions of true slip was determined both in one scan field and with multiple scans (200-400) in one place. The measurements were carried out at room temperature and at a heating up to 120°C. It is shown that at heating up to 75-85°C, the friction coefficient of some coPUI decreases significantly. The same effect can be achieved also after 100 scans during multi-scan testing at 20°C.

1. Introduction

The surface properties of the polymer films play a decisive role in their applications especially if they used as coatings [1]. Atomic force microscopy (AFM) is an indispensable technique to study the various surface phenomena [1-4]. AFM allows not only evaluating the roughness of the surface, but also recognizing the different localized phases and revealing the distribution of friction forces over the surface in the micro- and nanoscale [5-7].

The aim of this work was to study the tribological characteristics of a series of novel multiblock (segmented) copolymers, containing rigid imide blocks and flexible blocks of polyurethanes – copoly(urethane-imide)s (coPUI)s and their nanocomposites, in microscale using AFM in contact mode. The study included the friction coefficient measurement at the temperature range from 20 to 120 ^{O}C and multipass friction force determination over 400 scanning areas.

2. Experimental details

For synthesis of new multiblock coPUIs, the 1,3-bis (3’,4-dicarboxy-phenoxy) benzene (dianhydride R) and 4,4’-bis-(4’-aminophenoxy)diphenylsulfone (diamine SOD-*pair*) with various ester fragments, as well as an aliphatic copolyester (Alt), secondarily terminated with 2,4-tolylene diisocyanate, were used [8]. The chemical structures of the coPUIs are presented in Figures 1 and 2.

The nanocomposite films based on them containing from 1 to 10 wt. % carbon nanofillers of different morphology – single-walled carbon nanotubes (SWCNT), carbon nanofibers (VGCF), graphene, and WS_2 and WSe_2 nanoparticles were prepared. The studied samples are listed in Table 1.

IOP Conf. Series: Materials Science and Engineering **256** (2017) 012022 doi:10.1088/1757-899X/256/1/012022

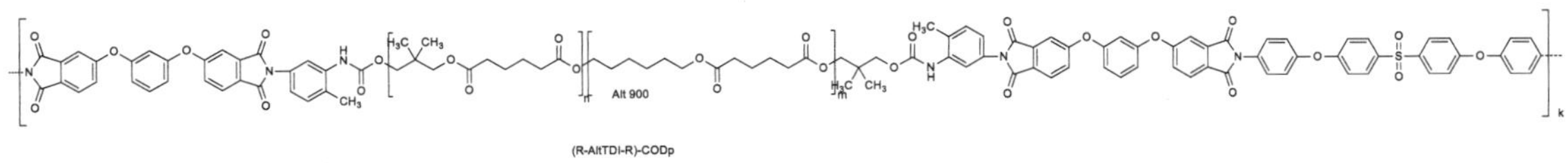

Figure 1. Chemical structure of coPUI (R-AltTDI-R)SOD*p* matrix.

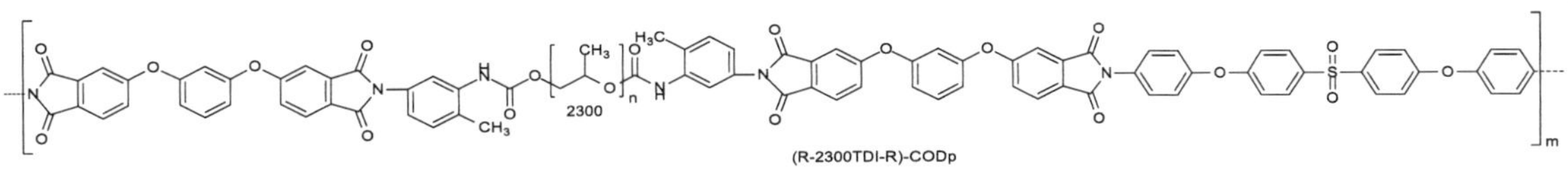

Figure 2. Chemical structure of coPUI (R-2300TDI-R)SOD*p* matrix.

The microscale roughness, friction coefficient, and friction force of the samples were investigated using AFM device NT-206 (produced by MTM Belarus). The surface morphology was studied using standard CSC38 silicon probe with the radius of tip curvature less than 10 nm and stiffness of cantilever 0.08 N/m produced by MikroMash (Estonia). The surface roughness was investigated in scanning areas 2x2, 5x5, and 10x10 μm^2 (Table 2).

The force and friction coefficient were measured using 20x20 μm^2 scanning areas with silicon probe NSC11 of V-shaped type with stiffness 3 N/m at the moving speed 1.55 µm/s. The tip radius was specially blunted by pre-scanning to curvature 100 nm. Such preparation prevents the probe from further blunting and keeps the contact condition stable. C_{fr} is calculated from a ratio of the friction force between two bodies and the normal force pressing them together; their measurement is described in [2]. To measure the friction force, the AFM probe scans the surface in so-called two pass method. During the first pass, the probe goes forward along the surface. On the second pass, it moves according to the first line, which suits the relief. During the forward and backward ways the friction forces from the surface influence on the tip, which causes the torsion of the cantilever besides its deflection in the normal direction (angle α in Figure 3). The angle of this twist depends on the value of the friction force. To calculate the effect of frictional forces more precisely, the angle of twist, when the probe is moved in the forward and backward directions, is divided in half.

Table 1. Samples of coPUIs and nanocomposites based on them investigated in this work

Matrix	Matrix symbol	Additive	Additive content, %	Composite symbol
(R-2300TDI-R)SOD*p*	50	-	-	50
		VGCF (nanofiber)	1	50 + 1% VGCF
		SWCNT (single-walled carbon nanotubes)	1	50 + 1% SWCNT
		Graphene	1	50 + 1% Graphene
		WS_2	3	50 + 3% WS_2
		WSe_2	1	50 + 1% WSe_2
(R-AltTDI-R)SOD*p*	45	-	-	45
		WS_2	10	45 + 10% WS_2
		WSe_2	10	45 + 10% WSe_2
		VGCF (nanofiber)	1	45 + 1% VGCF
		SWCNT (single-walled carbon nanotubes)	1	45+ 1% SWCNT
		Graphene	10	45 + 10% Graphene

Table 2. Roughness of the studied samples measured using AFM

Composite symbol	Side	Roughness, nm					
		Scanning area 10x10 μm²		Scanning area 8x8 μm²		Scanning area 5x5 μm²	
		Ra	Rq	Ra	Rq	Ra	Rq
50 + 1% Graphene	1	–	–	21.98	33.54	3.23	5.09
	2	–	–	20.51	26.04	18.14	23.09
50 + 1% SWCNT	1	–	–	43.73	56.40	21.69	26.11
	2	–	–	44.62	53.19	21.42	26.46
50 + 1% VGCF	1	6.21	12.91	–	–	3.40	4.40
	2	47.74	62.14	–	–	26.69	34.54
45 + 1% VGCF	1	8.16	11.84	–	–	7.54	10.6
	2	22.82	28.71	–	–	19.39	24.21
45 + 10% Graphene	1	–	–	87.61	110.2	148.5	177.0
	2	–	–	46.93	58.46	25.94	32.99
45+ 1% SWCNT	1	–	–	21.24	26.29	13.81	16.95
	2	–	–	76.13	95.96	72.40	83.28
45	1	4.53	5.71	–	–	2.69	3.36
	2	–	–	3.948	4.48	–	–

C_{fr} was determined in several regimes: for one scan field for one line in the forward and backward directions, and for the multiple scans (200-400) in one place. The measurements were carried out both at room temperature and at 120°C. The heating was carried out directly during the scanning process using a thermoplatform placed on the AFM table. The thermoplatform is equipped with heaters and a feedback system with sensors, which allows controlling the heating and automatic reducing the voltage, if necessary, to maintain the fixed temperature.

3. Results and discussion

Typical images of the coPUI matrices and nanocomposite surface are presented in Figure 4. According to AFM images, the differences in morphology of the films surface between two sides of the films can be clearly seen. It can be concluded that their structure and morphology are extremely sensitive to the substrate surface. Hence, the effect of additives influence on the films structure is much greater than confirmed by AFM images.

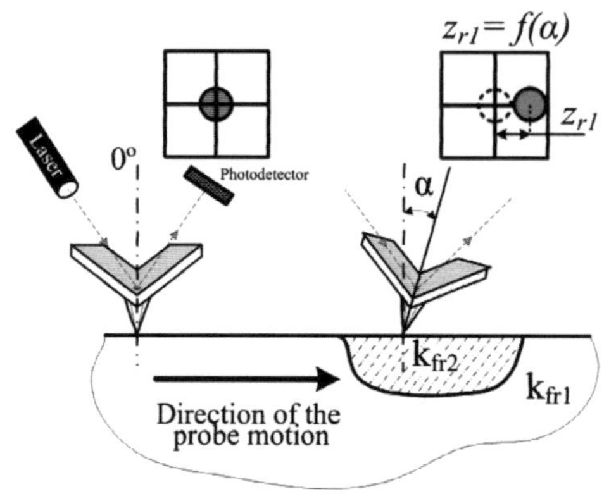

Figure 3. The scheme of cantilever twist under the action of friction forces.

(a) (b) (c)

Figure 4. Three-dimentional images: (a) neat (R-AltTDI-R)SOD*p* matrix; (b) R-AltTDI-R SOD*p* matrix with WS_2 nanoparticles; (c) coPUI (R-2300TDI-R)SOD*p* matrix with WS_2 nanoparticles. Scanned area 2x2 μm^2.

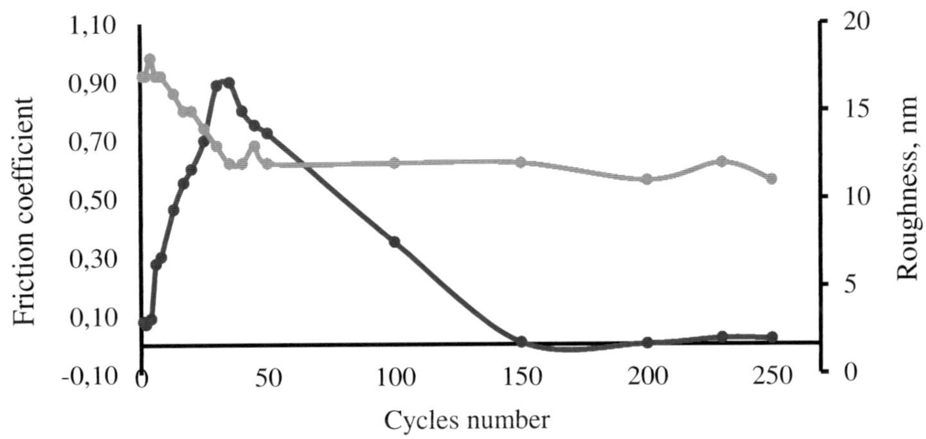

(a)

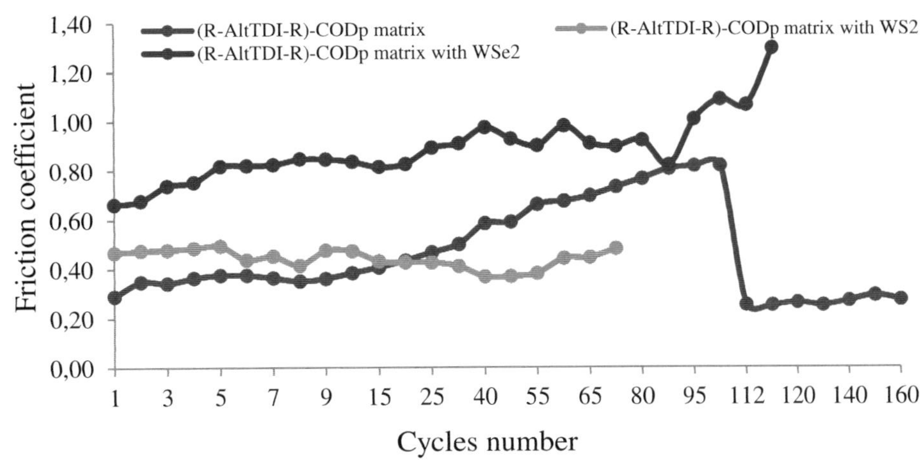

(b)

Figure 5. Dependence of the friction coefficient on the number of scanning cycles during multipass AFM testing: (a) 20°C, F normal 175 nN, friction forces 10-160 nN. Green curve – roughness, blue curve – C_{fr}; (b) (R-AltTDI-R)SOD*p* matrix (30°C, F normal 45 nN, friction forces 14-12 nN), with WS_2 (20°C and F normal 48 nN, friction forces 45-65 nN), with WSe_2 (30°C, F normal 54 nN, friction forces 30-60 nN).

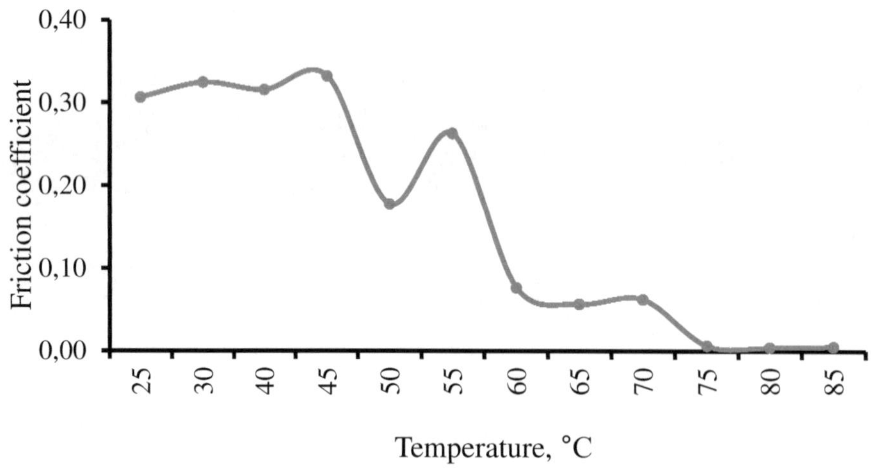

Figure 6. Dependence of the friction coefficient on the temperature during tribological AFM testing of (R-AltTDI-R)SOD*p* matrix with WS_2 with thermal platform under heating from 25 to 85°C (F normal = 69 nN, friction forces = 1-60 nN).

The significant difference in the topography of the opposite films surfaces leads to a significant difference in roughness (Table 2). (R-2300TDI-R)SOD*p* + 1% Graphene and (R-2300TDI-R)SOD*p* + 1% SWCNT films have the close roughness values on the opposite sides. The roughness of (R-2300TDI-R)SOD*p* + 1% VGCF films and all films of nanocomposites based on (R-AltTDI-R)SOD*p* matrix differ by two or three times. Thus, additives and substrates significantly influence on the structure formation of these PI films and on the properties of their surfaces.

C_{fr} of matrices without additives differ threefold: 0.164-0.167 for (R-2300TDI-R)SOD*p* and 0.056-0.065 for (R-AltTDI-R)SOD*p*. The addition of VGCF reduces C_{fr} of (R-2300TDI-R)SOD*p* matrix more than twice to 0.061-0.074 and leaves C_{fr} of (R-AltTDI-R)SOD*p* matrix at the low level about 0.066-0.071.The addition of SWCNT practically leaves C_{fr} of both matrices on the level of C_{fr} of pure matrices. The addition of graphene increases C_{fr} of both matrices, especially that of (R-2300TDI-R)SOD*p*. Thus, only the addition of VGCF to (R-2300TDI-R) SOD*p* matrix improves the tribological properties of the surface. The excess of the quantity necessary for the modification, as in the case of coPUI with 10% graphene, can substantially worsen the surface properties.

In general, WS_2 and WSe_2 nanoparticles degrade C_{fr} of the (R-2300TDI-R)SOD*p* and (R-AltTDI-R)SOD*p* matrices due to a significant increase of roughness. WSe_2 decreases twice the C_{fr} of (R-AltTDI-R)SOD*p* matrix surface formed on glass.

The dependences of roughness and C_{fr} value on the number of scans during multipass tribological AFM testing of (R-AltTDI-R)-CODp matrix are presented on Figure 5a. This measurement was performed at 20°C and normal load about 175 nN. Friction forces ranged from 10 to 160 nN. Each point on the C_{fr} graph (blue) and the roughness graph (green) is obtained from a separate AFM sequential scanning field (Fig. 5). At the initial stage of C_{fr} graph up to 50 cycles, the growth of C_{fr} to 0.9 was determined and the roughness simultaneously decreased. When the probe moves along the surface, the roughness is smoothed out. The low initial value of C_{fr} 0.1 can be attributed to developed roughness when the probe contacts with surface along the tops of the irregularities. In this case, the frictional forces are lower than those for the contact of the smoothed sample with entire surface of the probe. The energy should be spent on the process of smoothing and it is necessary to apply a greater lateral force for further movement [2]. Therefore, K_{tr} grows on this site. In the process of friction, the irregularities are smoothed out and after reaching the minimum of roughness, C_{fr} decreases over the next 100 cycles to a stable value of 0.05. And then the mechanism of elastic non-wear friction works.

When the same multipass tests are repeated at 30°C, the probe-sample contact conditions change, which raises the initial value of C_{fr} to 0.3 instead of 0.1 at 20°C (Fig. 5b). The normal load was about 116 nN and measured friction forces in a range of 30-92 nN. C_{fr} is stable at the beginning during the first 10 cycles and then it increases smoothly. After reaching 0.8 at 110 cycles as well as at 20°C the curve decreases to a stable level 0.25. The testing at 30°C places peculiarities on the C_{fr} graph: growth is smoother and decrease is faster. The results of friction change during multipass tribological AFM testing at 30°C of (R-AltTDI-R)SOD*p* matrix with WS_2 and WSe_2 nanoparticles are shown on Figure 5b. As for pure (R-AltTDI-R)SOD*p* matrix the results of testing at higher temperature show the increased C_{fr} in comparison with the testing at 30°C. The addition of WS_2 nanoparticles keeps C_{fr} stable on the level 0.45. The addition of WSe_2 nanoparticles leads to C_{fr} increase.

The influence of heating on C_{fr} of (R-AltTDI-R)SOD*p* with WSe_2 is shown in Figure 6. C_{fr} is stable till 45°C. Then it dramatically decreases at 60°C and keeps stable till 70°C, while at 75°C it decreases. Comparing the similar values of C_{fr} of (R-AltTDI-R)SOD*p* matrix determined for «multipass scan» and «one scan under heating», it should be noticed that the influence of heating from 75 to 85°C on the surface properties is similar to the influence of 200-400 testing scans at 20°C. This fact allows us to conclude that heating in this micro contact is equal to 75°C. These results show that synthesized coPUI films are very promising for tribological applications and their properties can be controlled both by temperature and mechanical action.

4. Conclusion

The tribological properties of new thermoplastic elastomers based on (R-AltTDI-R)SOD*p* and (R-2300TDI-R)SOD*p* coPUI matrices modified with carbon materials (nanotubes, graphene, and fibers) were tested in microscale by atomic force microscopy.

Comparing the similar values of C_{fr} of (R-AltTDI-R)SOD*p* matrix determined in regimes «multipass scan» and «one scan under heating», it should be noticed that the influence of heating in the range of 75-85 °C on the surface properties is similar to the influence of 200-400 scans of multipass testing at 20 °C. This fact allows making a conclusion that there is a heating in this micro contact equal to 75 °C. These results show that synthesized coPUI films and nanocomposites based on them are very promising materials for tribological applications and their properties can be controlled both by temperature and by mechanical action.

It is shown that the addition of carbon materials to the studied films significantly changes their friction coefficient due to extreme sensitivity of coPUI matrices to the synthesis conditions, solvent evaporation rate, and surface effects. Thus, the friction coefficient can be controlled by the addition as well.

Therefore, it is shown that atomic force microscopy is an excellent research tool in the field of tribology in nanoscale.

Acknowledgments

This research was supported by the grants of Belarussian Republican Foundation for Fundamental Research BRFFR No. F16R-142 and Russian Foundation for Basic Research RFBR No. 16-53-00178.

References

[1] Al-Ajaj I and Kareem A 2016 *Materials Science - Poland* **34** 132-6
[2] Bhushan B, Israelachvili J N and Landman U 1994 *Nature* **374** 607-16
[3] Chizhik S A, Rymuza Z, Chikunov V V, Jarzabek D and Kuznetsova T 2007 *Recent advances in mechatronics* Ed.: R. Jabłoński [et al.] (Berlin, Heidelberg: Springer) 541-5
[4] Anishchik V, Uglov V, Kuleshov A, Filipp A, Rusalsky D, Astashynskaya M, Samtsov M, Kuznetsova T, Thiery F and Pauleau Y 2005 *Thin Solid Films* **482** 248-52
[5] Warcholinski B et. al. 2017 *Surf. Coat. Technol.* **309** 920-30
[6] Kuznetsova T, Zubar T, Chizhik S, Gilewicz A, Lupicka O and Warcholinski B 2016 *J. Mater. Eng. Perform.***25** 5450-9
[7] Warcholinski B, Gilewicz A, Kuznetsova T A, Zubar T I, Chizhik S A, Abetkovskaia S O and Lapitskaya V A 2017 *Surf. Coat. Technol.* **319** 117-28
[8] Didenko A, Yudin V, Smirnova V, Gofman I, Popova E, Elokhovskii V, Svetlichnyi V and Kudryavtsev V 2014 *Mater., Methods and Technol.* **8** 31-40

IOP Conf. Series: Materials Science and Engineering **256** (2017) 012023 doi:10.1088/1757-899X/256/1/012023

Profiling of nanostructures by scratching probe nanolithography and plasma chemical etching

R V Tominov[*], **V V Bespoludin, V S Klimin, V A Smirnov and O A Ageev**

Southern Federal University, Institute of nanotechnologies, electronics and equipment engineering, Research & Education Center «Nanotechnologies», Taganrog, Russia

[*]roman.tominov@gmail.com

Abstract. This work presents the results of the experimental investigations of nanostructure profiling using scratching probe nanolithography and plasma chemical etching. It is shown, that combination of these methods allows us to profile the silicon surface with height from approximately 7,8 to 38 nm and diameter from 670 to 123 nm.

1. Introduction

Reducing the size of the elements of integrated circuits (IC) is one of the main tasks of micro- and nanoelectronics [1-5]. The modern level of technology requires the formation of the elements with size below 100 nm. The photolithography cannot be used for creation of such structures due to its resolution limit. Therefore, there is a need to develop new techniques for production of IC elements on a nanometer scale [5, 6]. Scratching probe nanolithography of atomic force microscopy (AFM) is one of the promising methods on this way [7]. The method is a mechanical modification of the photoresist surface using a probe tip [8-18]. After the structures profiling is done, plasma-chemical etching of the substrate can be carried out [13, 14]. The photoresist layer plays the role of a mask.

The aim of this work is investigation of profiling of nanostructures using scratching probe nanolithography and plasma-chemical etching.

2. Experiment details

The diagram of nanostructures formation on the Si surface is shown in Figure 1. A positive photoresist FP-383 mixed with a diluent RPF-383 in the ratio 1:14 was used as an experimental sample. This solution was applied to the silicon substrate by centrifugation at a rotational speed of 5500 rpm (Fig. 1a). Then the sample was dried at 90°C for 2 minutes. The film thickness was 20±2 nm.

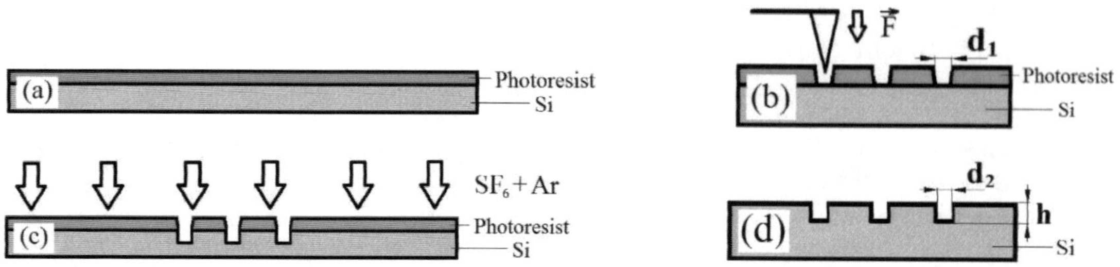

Figure 1. Diagram of nanostructure formation on Si: (a) deposition of photoresist; nanostructures formation: (b) on photoresist by scratching probe nanolithography; (c) on Si surface by plasma-chemical etching; (d) on Si substrate after photoresist removing.

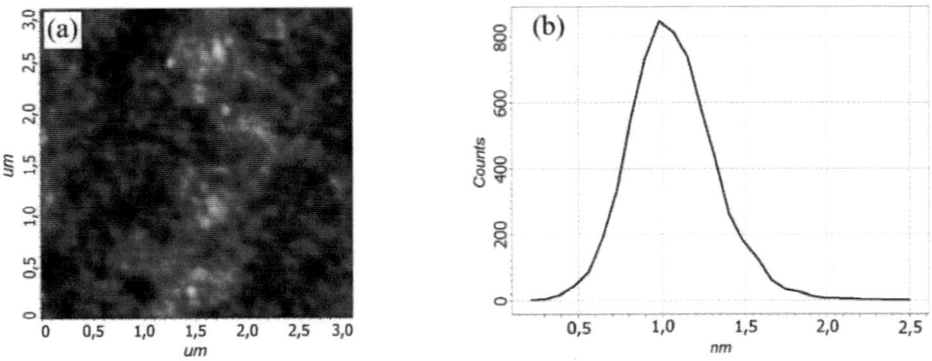

Figure 2. Photoresist surface morphology: (a) AFM image; (b) roughness histogram.

Scratching probe nanolithography was carried out at the Ntegra probe nanolaboratory (NT-MDT, Russia) in contact mode AFM using a cantilever NSG-11. As a result, arrays 3×3 μm^2 of profiled nanoscale structures with diameters d_1 from 60±4 nm to 436±27 nm were formed in different regions of the photoresist surface at $F = 3$ μN (Fig. 1b).

For experimental studies of plasma interaction with Si under inductively coupled plasma, a cylindrical flow-through plasma chemical reactor was used. The pressure of a mixture of plasma-forming gas was 2 Pa. Plasma etching of the samples was carried out in STE ICPe68 (SemiTEq JSC, Russia). Nanostructures were formed in combined gas discharge plasma with following parameters: power source inductively coupled plasma W_{ICP}=300 W; power capacitive plasma source W_{RIE}=35 W; bias voltage U_{bias} = 73 V. For etching, a working mixture of gases Ar:SF_6 in the ratio of 10:1 was used; etching time was about 1 min at 23°C (Fig. 1c). After that photoresist residues were removed in dimethylformamide (Fig. 1d). AFM images of nanostructures were obtained in semicontact mode to define diameters d_2 and height h (Fig. 1d). Images were analysed using Image Analysis 2.0 software.

3. Results and discussion

The surface was analyzed before scratching probe process using AFM (Fig. 2a) and it was shown that surface roughness was 0.97±0.61 nm (Fig. 2b). This implies that the photoresist film can be used for scratching probe nanolithography.

It is important that the height (Fig. 1b) of the formed nanostructures on the photoresist surface must be equal to photoresist thickness. Therefore, after formation of the profiled nanostructures on the photoresist surface the AFM images were obtained to examine the nanostructure height (Fig. 3a). It is shown that structures height 20±2 nm (Fig. 3b) is equal to the thickness of the photoresist film.

After scratching probe nanolithography, plasma chemical etching and photoresist removing, AFM images of nanostructures on Si surface were obtained to examine their heights and diameters (Fig. 4).

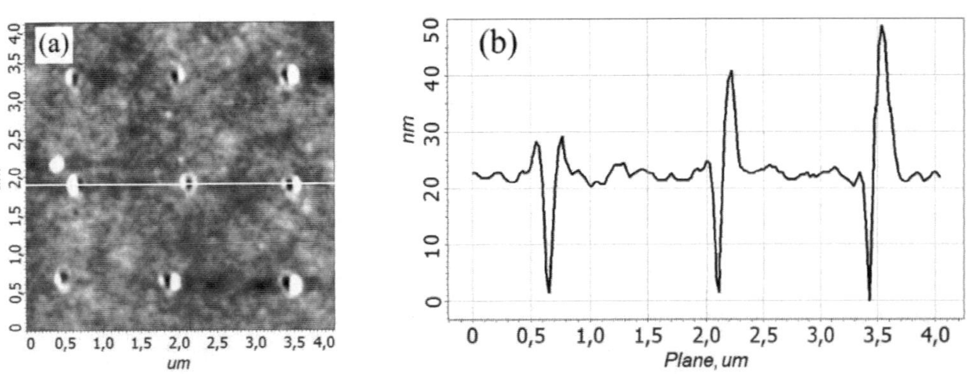

Figure 3. Profiled nanostructures formed on photoresist surface using scratching probe nanolithography: (a) AFM image; (b) profile along white line on (a).

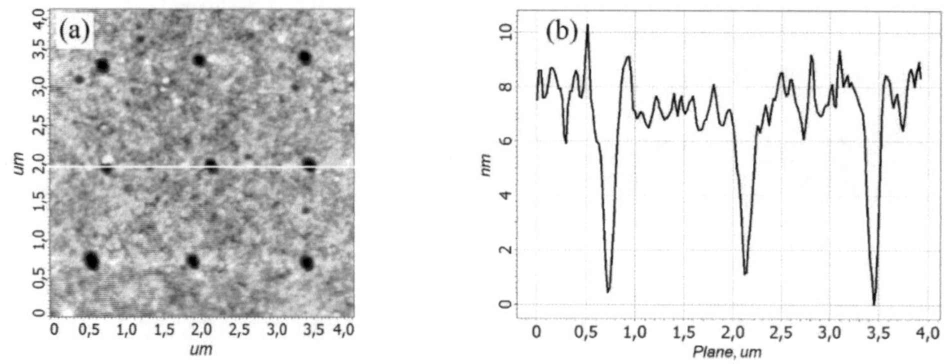

Figure 4. Profiled nanoscale structures on Si surface after plasma-chemical etching and photoresist removing: (a) AFM image; (b) profile along white line on (a).

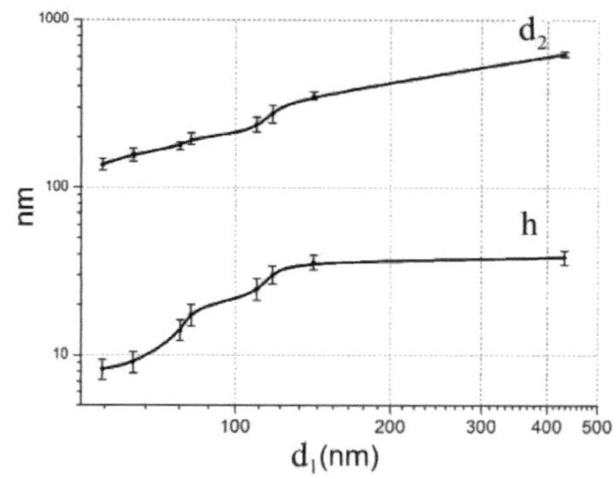

Figure 5. Dependence of the diameter d_1 of nanostructures at the photoresist surface on the height h and diameter d_2 of nanostructures at Si surface.

Dependence of the height h of nanostructures and the diameter d_2 on the diameter d_1 were plotted (Fig. 5). It is shown that increase of d_1 from 68±3 to 428±48 nm leads to increase of h from 8±1 to 38±5 nm and increase of the diameter d_2 from 123±17 to 672±48 nm.

4. Conclusion

In summary, we have investigated regimes of profiling of nanostructures using scratching probe nanolithography and plasma chemical etching. It has been shown that the combination of these methods allows profiling the silicon surface with height h from approximately 7.8 to 38 nm and diameter d_2 from 670 to 123 nm. The results can be used in the micro- and nanoelectronics elements manufacturing using probe nanotechnology.

Acknowledgments

The study was supported by the RFBR (projects 16-29-14023 ofi_m, 16-32-00069 mol_a) and Southern Federal University (internal grants 07/2017-02, 07/2017-26). The results were obtained using the equipment of Research and Education Center and the Center for Collective Use "Nanotechnologies" of Southern Federal University.

References

[1] Akhavan O and Abdolaha M 2009 *Applied Surface Science* **255** 3513
[2] Tang Q, Shi S Q and Zhou L 2004 *J. Nanosci Nanotechnol* **4** 948
[3] Garcia R, Armin W and Riedo E 2014 *Nature nanotechnology* **9** 577
[4] Ageev O A, Smirnov V A, Solodovnik M S, Rukomoikin A V and Avilov V I 2012 *Semiconductors* **46** 1616

[5] Avilov V I, Ageev O A, Smirnov V A, Solodovnik M S and Tsukanova O G 2015 *Nanotechnologies in Russia* **10** 214
[6] Ageev O A, Alyab'eva N I, Konoplev B G, Polyakov V V and Smirnov V A 2010 *Semiconductors* **44** 1703
[7] Avilov V I, Ageev O A, Kolomiitsev A S, Konoplev B G, Smirnov V A and Tsukanova O G 2013 *Semiconductors* **106** 50
[8] Bukharaeva A A, Bizyaeva D A, Nurgazizova N I and Khanipova T F 2012 *Russian Microelectronics* **41** 90
[9] Bizyaev D A, Bukharaev A A, Nurgazizov N I and Khanipov T F 2012 *Technical Physics Letters* **38** 8
[10] Klimin V S, Solodovnik M S, Smirnov V A, Eskov, A.V., Tominov R V and Ageev O A 2016 *Proceedings of SPIE* **10224** 102241Z
[11] Sung I H and Kim D E 2005 *Applied Surface Science* **239** 209
[12] Heyde M, Rademann K, Cappella B, Geuss M, Sturm H, Spangenberg H and Niehus H 2001 *Review of Scientific Instruments* **72** 136
[13] Chen Y J, Hsu J H and Lin H N 2005 *Nanotechnology* **16** 1112
[14] Ageev O A, Klimin V S, Solodovnik M S, Eskov A V and Krasnoborodko S Y 2016 *J. of Physics: Conference Series* **741** 012178
[15] Sumomog T, Endo T, Kuwahara K and Kaneko R 1995 *J. of Vacuum Science & Technology* **13** 1257
[16] Acikgoz C, Hempenius M A, Huskens J and Vancso G J 2011 *Macromolecular Nanotechnology* **47** 2033
[17] Bouchiat V and Esteve D 1996 *Appl. Phys. Lett.* **69** 3098
[18] Burnham N A and Colton R J 1989 *J. of Vacuum Science & Technology* **7** 2906

Piezoelectric effect in non-uniform strained carbon nanotubes

M V Ilina*, Yu F Blinov, O I Ilin, N N Rudyk and O A Ageev

Southern Federal University, Research and Education Center "Nanotechnologies", Taganrog, 347922, Russia

*mailina@sfedu.ru

Abstract. The piezoelectric effect in non-uniform strained carbon nanotubes (CNTs) has been studied. It is shown that the magnitude of strained CNTs surface potential depends on a strain value. It is established that the resistance of CNT also depends on the strain and internal electric field, which leads to the hysteresis in the current-voltage characteristics. Analysis of experimental studies of the non-uniform strained CNT with a diameter of 92 nm and a height of 2.1 μm allowed us to estimate the piezoelectric coefficient 0.107 ± 0.032 C/m^2.

1. Introduction

The transition of electronics into the nanoscale area has led to the fact that the surface piezoelectric and flexoelectric effects have a significant impact on the electrical properties of the nanostructures [1, 2]. These effects were not taken into account in bulk materials because of the small values of the surface-to-volume ratio and the strain gradients. However, in the nanostructures the surface-to-volume ratio and the strain gradients can reach high values [1].

A new science direction – nanopiezotronics, which studies the flexo- and piezoelectric properties of nanostructures, has developed [3]. The fundamental principles of nanopiezotronics were formulated less than ten years ago [3]. However, the search for materials for its practical implementation continues at present. Currently, there are reports that carbon nanostructures can exhibit flexo- and piezoelectric properties [4-6], which opens a wide field for their application in nanopiezotronics and nanoelectronics at high strength and elasticity.

The aim of this work is to study a piezoelectric effect in non-uniform strained carbon nanotubes (CNTs) and to determine their piezoelectric coefficient using scanning probe microscope.

2. Experimental studies

2.1. Experimental sample

An array of vertically aligned CNTs grown by plasma chemical vapor deposition using nanotechnology complex NANOFAB NTC-9 (NT-MDT, Russia) was studied. CNTs were grown by a vertex mechanism using Ni catalyst particles. Figure 1a shows the sample image obtained by scanning electron microscope (SEM) Nova NanoLab 600 (FEI, Netherlands). Structural analysis by Raman spectroscopy showed presence of D- and G-mode characteristic of carbon nanotubes and absence of RBM mode in the range of 0-200 cm^{-1}, indicating that they were multi-walled. Analysis of transmission electron microscopy (TEM) image showed that the experimental sample represented multi-walled carbon nanotubes with bamboo-like defects (Fig. 1b).

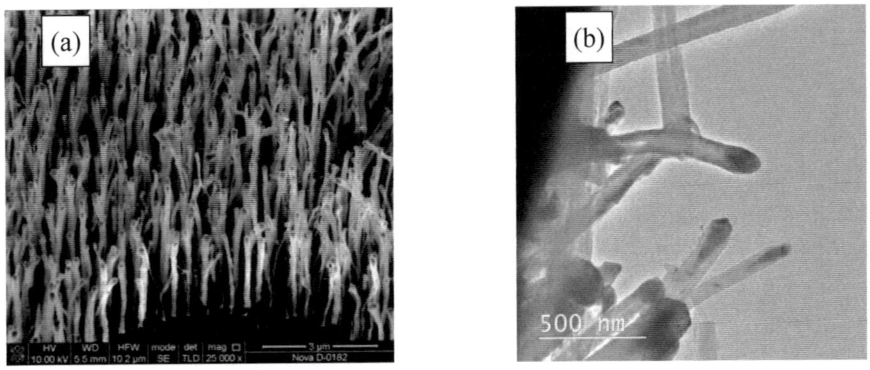

Figure 1. The experimental sample of CNT: SEM (a) and TEM (b) images.

2.2. Methods

To study the piezoelectric effect in CNTs, a scanning of the array surface by atomic force microscopy (AFM) in a tapping mode was carried out on the first pass (Fig. 2a). As a result, the CNT bundles having non-uniform bending strain were formed by van der Waals forces [7]. During the second pass, the surface potential of the strained CNT bundles at a distance of 12 nm from the array surface was measured. The surface potential of strained CNTs was studied by Kelvin probe mode using the probe nanolaboratory Ntegra (NT-MDT, Russia). The NSG11/Pt cantilever was used. The image of the distribution of the surface potential of the CNT bundles is shown in Figure 2b.

To determine the piezoelectric coefficient of CNT, a controlled non-uniform strain was formed in it and the current-voltage characteristics (CVCs) were measured. Controllable non-uniform strain in CNTs was formed under the action of a local electric field in scanning tunnel spectroscopy mode [6]. The strain value of CNT from 0.2 to 3.0 nm was controlled by feedback system of the scanning tunneling microscope (STM). The strain of CNT was maintained by van der Waals forces between STM probe and nanotube top after removing the external electric field [8]. CVCs of the strained CNTs were measured by applying a saw tooth voltage pulse with amplitude of 8 V and duration of 1 s. Upper electrode was a tungsten STM probe. Lower electrode was conductive layer on a substrate with CNTs.

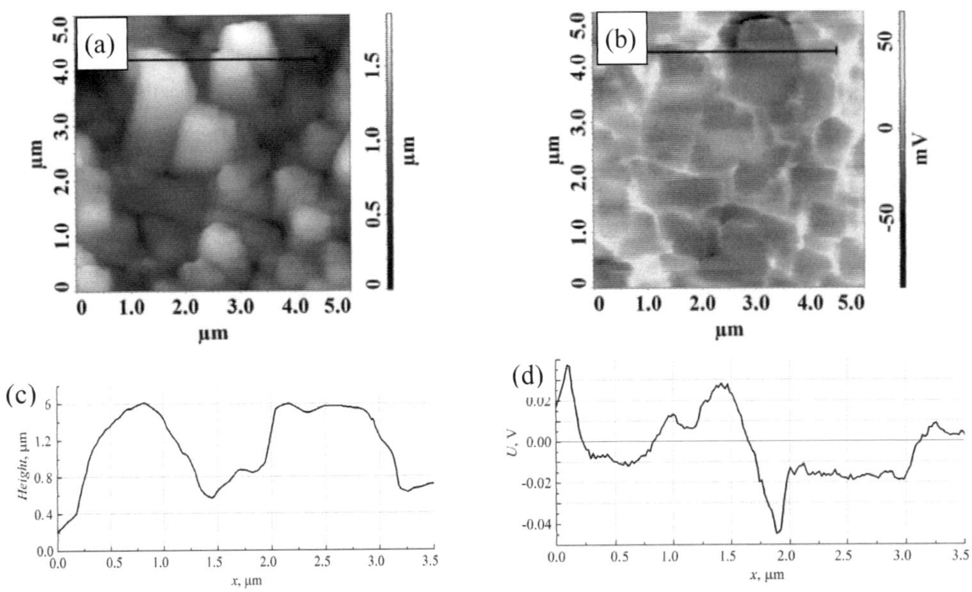

Figure 2. Investigation of the surface potential of non-uniform strained CNTs forming a bundles: the surface topology (a), the potential distribution (b), and sections along the lines (c, d).

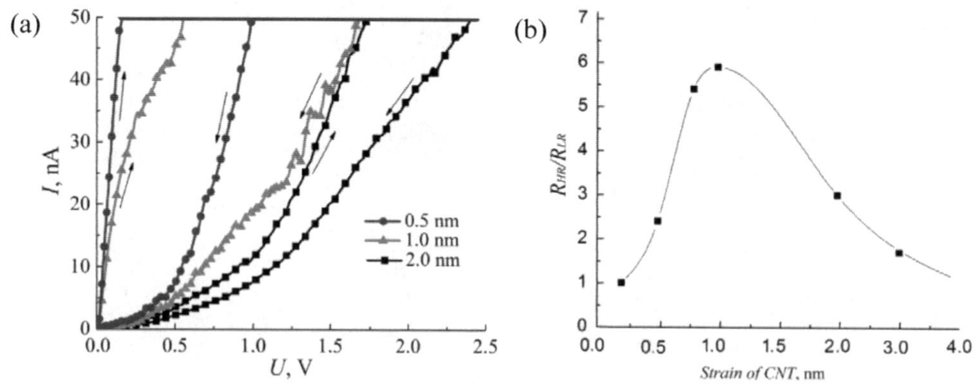

Figure 3. The electrical characteristics of the CNT at different values of its strain: (a) CVCs, (b) ratio of the resistances in the high- and low-resistance states by reading voltage at 0.2 V.

3. Results and discussion

Analysis of the AFM images shows that a negative surface potential of up to -42 mV is observed at the tops of the CNT bundles up to 500 nm in diameter (Fig. 2c). A positive potential of up to 40 mV is observed near the bundles bottom (Fig. 2d). Thus, the magnitude of the surface potential depends on the diameter of the bundle and the strain value of nanotube in it (Fig. 2c, d). The potential at the tops of individual nanotubes is close to zero (Fig. 2d). The presence of a surface potential on bundles of the strained CNTs agrees with the results of the previous studies [9, 10] and confirms the possibility of the piezoelectric effect in carbon nanotubes.

The measuring of CVCs of the non-uniform strained CNT with a diameter of 92 nm and a length of 2.2 μm shows that the largest loop is observed for the CNT having a strain at 1 nm (Fig. 3a) due to the fact that for a given strain an internal electric field in the CNT is formed; it is completely redistributed when the voltage (8 V) is applied due to the piezoelectric effect [6]. As a result, the resistance of the CNT in the low-resistance state (R_{LR}) is determined by resistivity of the nanotube.

The hysteresis of CVCs decreases in the deviation from this strain value (Fig. 3a). This is due to the fact that the internal electric field of the CNT becomes insufficient to form a high-resistance state (R_{HR}) at a lower strain of the nanotube (less than 1 nm). An increase in the CNT strain value, on the contrary, led to an increase in the internal electric field, as the resistance of the CNT in the low-resistance state increased sharply (Fig. 3a). Thus, the ratio of the resistances of the CNT in the high-resistance and low-resistance states (R_{HR}/R_{LR}) was maximal at a strain equal to 1 nm and decreased as it increased or decreased (Fig. 3b).

We can conclude that the internal electric field of CNT with a diameter of 92 nm and a length of 2.1 μm is close to zero, when its strain is 1 nm and the voltage of 8 V is applied. Therefore, the intrinsic resistance of the CNT is about 25 MΩ (Fig. 3a). Then the resistance R_{LR} of the CNT increases from 202 to 285.7 MΩ with an increase in strain from 1.5 to 2.0 nm, respectively. If we assume that the increase in the CNT resistance is due to the growth of its internal electric field by reason of the appearance of a piezoelectric effect, then the piezoelectric coefficient of the CNT with a diameter of 92 nm and a length of 2.1 μm is 0.107±0.032 C/m^2. Our previous calculation using the classical equations of the piezoelectric effect showed a lower value of the piezoelectric coefficient of the CNT, which was 0.048 C/m^2 [6]. The difference in the obtained values may be due to the non-uniform strain of the CNT and the increase in the internal electric field of the nanotube due to the manifestation of the flexoelectric effect in it [4, 5].

4. Conclusion

In summary, the obtained experimental results confirm the manifestation of the piezoelectric effect in multi-walled carbon nanotubes. It is shown that under the action of mechanical stresses the CNTs are polarized and the internal electric field is formed, which leads to a change in their resistance. Analysis of experimental studies of the non-uniform strained CNT allowed us to estimate the value of the piezoelectric coefficient 0.107 ± 0.032 C/m^2.

The obtained results can be used in the development and the fabrication of nanopiezotronics and nanoelectronics devices based on carbon nanotubes.

Acknowledgments
The study was financially supported by the Russian Foundation for Basic Research (project No.16-29-14023 ofi_m) and Internal grant of the Southern Federal University (project No. VnGr-07/2017-26). The results were obtained using the equipment of Research and Education Center and the Center for collective use "Nanotechnologies" of Southern Federal University.

References

[1] Zubko P, Catalan G and Tagantsev A K 2013 *Annu. Rev. Mater. Res.* **43** 387-421
[2] Chandratre S and Sharma P 2012 *Appl. Phys. Lett.* **100** 023114
[3] Wang Z L 2007 *Adv. Mater.* **19** 889-892
[4] Kvashnin A G, Sorokin P B and Yakobson B I 2015 *J. Phys. Chem. Lett.* **6** 2740-4
[5] Kundalwal S I, Meguid S A and Weng G J 2017 *Carbon* **117** 462-72
[6] Il'ina M V, Il'in O I, Blinov Yu F, Smirnov V A, Kolomiytsev A S, Fedotov A A, Konoplev B G and Ageev O A 2017 *Carbon* **123** 514-24
[7] Ageev O A, Il'in O I, Rubashkina M V, Smirnov V A, Tsukanova O G and Fedotov A A 2015 *Tech. Phys.* **60** 1044-50
[8] Ageev O A, Blinov Yu F, Il'ina M V, Il'in O I, Smirnov V A and Tsukanova O G 2016 *Phys. Solid State* **58** 309-14
[9] Ageev O A, Blinov Yu F, Ilina M V, Ilin O I and Smirnov V A 2016 *J. Phys. Conf. Ser.* **741** 012168
[10] Wang X, Tian H, Xie W, Shu Y, Mi W T, Mohammad M A, Xie Q Y, Yang Y, Xu J B and Ren T L 2015 *NPG Asia Materials* **7** e154

IOP Conf. Series: Materials Science and Engineering **256** (2017) 012025 doi:10.1088/1757-899X/256/1/012025

In situ visualization of domain structure evolution during field cooling in 0.67PMN-0.33PT single crystal

A D Ushakov[1,*], A A Esin[1], D S Chezganov[1], A P Turygin[1], A R Akhmatkhanov[1], Q Hu[1], L Sun[1], X Wei[2] and V Ya Shur[1]

[1] School of Natural Sciences and Mathematics, Ural Federal University, 620000 Ekaterinburg, Russia
[2] Electronic Materials Research Laboratory, Key Laboratory of the Ministry of Education & International Center for Dielectric Research, Xi'an Jiaotong University, 710049 Xi'an, China

[*]andrey.ushakov@urfu.ru

Abstract. The evolution of the domain structure during in-field cooling was *in situ* studied in [001]-cut single crystals of relaxor ferroelectric (1-x)$Pb(Mg_{1/3}Nb_{2/3})O_3$–x$PbTiO_3$ (PMN-PT) with x = 0.33 with maximum of dielectric permittivity at 150°C. The main stages of domain evolution have been separated. The visualization of the static as-grown and polarized domain structures with high spatial resolution by piezoresponse force microscopy and scanning electron microscopy allowed measuring the characteristic features of maze and needle-like domain structures.

1. Introduction

Relaxor ferroelectric single crystals, such as *(1-x)*$Pb(Mg_{1/3}Nb_{2/3})O_3$-*x*$PbTiO_3$ (PMN-PT) and $Pb(Zn_{1/3}Nb_{2/3})O_3$-$PbTiO_3$ (PZN-PT), have attracted extensive attention over the last 20 years due to their superior piezoelectric properties leaving far behind polycrystalline ceramics $Pb(Zr_xTi_{1-x})O_3$ (PZT) [1-3]. Comprehensive studies have been focused on the piezoelectric response of domain engineered crystals. PMN-PT is the most widely used relaxor single crystal due to its broad morphotropic phase boundary (MPB) and comparatively simple growth procedure [4]. Increase of PT concentration allows changing the phase of PMN-*x*PT at room temperature from the rhombohedral *R3m* (for x < 0.32) to intermediate monoclinic *Cm* (for 0.32 < x < 0.37) and tetragonal *P4mm* (for x > 0.37). At present, the investigators focus their interests on the domain engineered PMN-PT single crystals with various compositions near or far from MPB [5, 6]. However, the high resolution observation of the poling process in PMN-PT crystals is rarely reported.

The domain engineering in ferroelectric crystals represents the modern method of creation of the stable domain configurations with desired geometrical parameters for enhancement of the proper crystal properties [7, 8]. Artificially created high concentration of the domain walls leads to high piezoelectric properties in PMN-PT. The tailored domain structure with precise periodicity will allow realizing the quasi-phase matching effect in PMN-PT, thus obtaining the efficient frequency conversion of the coherent light.

We investigated the 0.67PMN-0.33PT single crystal, which is a typical MPB composition, in which coexistence of *Ma* and *Mc* phases has been confirmed experimentally [9]. We carried out an experimental observation of the domain structure evolution during the field cooling procedure using the optical microscopy and separated the main stages of the process. The static domain structures before and after field cooling were visualized using piezoresponse force microscopy (PFM) and scanning electron microscopy (SEM). The synergy of the combination of the used experimental methods allowed studying the process of domain structure evolution in multiaxial relaxor ferroelectric single crystals PMN-PT.

2. Experiment

Lead magnesium niobate-lead titanate $(1\text{-}x)Pb(Mg_{1/3}Nb_{2/3})O_3–xPbTiO_3$ (PMN-PT) single crystals with x = 0.33 corresponding to tetragonal phase at room temperature were grown from the melt by modified Bridgman technique [10]. The [001]-cut plates with dimensions 40×40×0.8 mm^3 were mechanochemically polished to optical grade using silica compound (SF1 suspension, Logitech Ltd.), then sliced to 5×1×0.8 mm^3 samples using DAAD 3220 dicing saw (Disco Co., Japan). The electrodes were formed on (100) sample surfaces.

The used experimental setup consisting of polarizing optical microscope Olympus BX51 (Olympus Corporation, Japan) with CCD-camera and heating stage THMSE600 (Linkam Scientific Instruments Ltd., UK) allowed recording the instantaneous domain patterns field treatment including: (a) zero field heating, (b) in-field annealing, and (c) field cooling. The procedure was realized by four consequent stages: (1) zero field heating to 200°C with rate 3°C/min, (2) in field annealing at constant temperature 200°C for 10 min, (3) slow cooling to 150°C with rate of 1°C/min, and (4) in-field cooling down to room temperature with rate of 10°C/min. The electric field with amplitude 300 V/mm was applied to the sample in [100] direction by voltage amplifier TREK 677B (Trek Inc., USA). Dielectric properties have been measured using impedance bridge QuadTech RLC 7600 (IET Labs Inc., USA). Frequency and amplitude of testing signal were 100 Hz and 1 V, respectively.

The shallow surface relief (about few nanometers) corresponding to domain patterns revealed by selective etching during mechanochemical polishing was visualized by SEM in detection of secondary electron mode using Merlin (Carl Zeiss, Germany) [11]. The static domain structures were visualized by scanning probe microscope NTEGRA Aura (NT-MDT, Russia) using PFM mode [12].

3. Results and discussion

Temperature and field profiles used for field cooling are shown in Figure 1a. We use a quite slow cooling rate through the Curie temperature to decrease internal stress and prevent the formation of non-180° domains. The temperature dependences of dielectric permittivity during heating and subsequent cooling measured for non-poled sample (Fig. 1b-c) showed a maximum around 150 °C. The single peak was observed during heating in the dielectric measurement of the non-poled sample. However, in the cooling stage one additional peak appeared around 55 °C. Generally, such two dielectric peaks, which correspond to the *Ma* to *Mc* and *Mc* to *C* phase transitions, respectively, are observed usually in the heating measurement of poled samples. The obtained temperature hysteresis with essentially higher value of dielectric permittivity for cooling stage could be attributed to input of the appeared mobile domain walls.

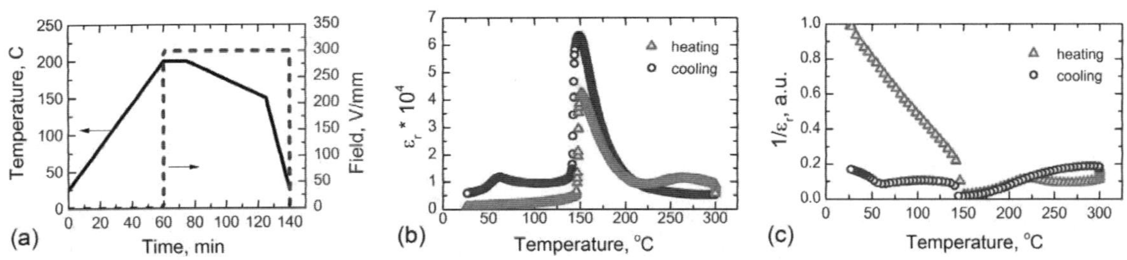

Figure 1. (a) Temperature and field profiles used for poling procedure. Temperature dependences of (b) dielectric permittivity and (c) reversed dielectric permittivity for heating and cooling stages.

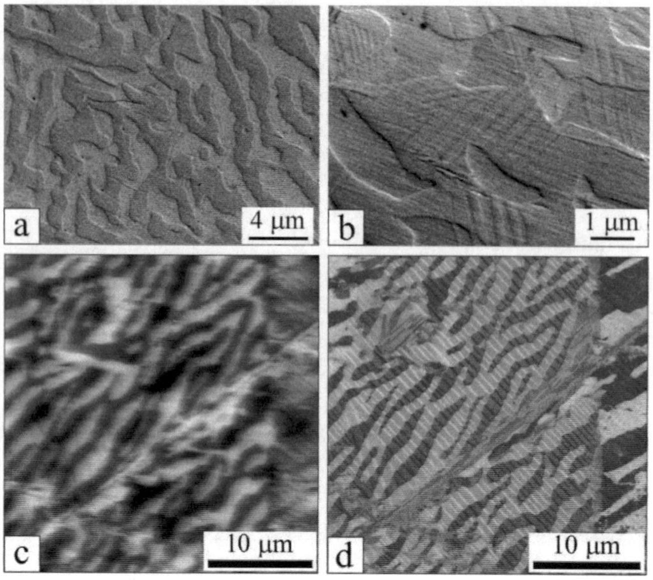

Figure 2. Images of as-grown domain structure after mechanochemical polishing: (a), (b) SEM and PFM: (c) topography and (d) amplitude.

3.1. As-grown domain structure
SEM images of the 0.67PMN-0.33PT single crystal etched during mechanochemical polishing are shown in Figure 2a,b. It is known that the etching allows revealing both in plane and out of plane domains. Maze adjacent domains and needle-like domains were revealed similar to that reported by Yu *et al.* [13]. The width of needle-like domains is about 150 nm and that of maze domains ranged from 0.5 to 3 microns. The formation of micro-sized domains is due to destruction of the long range ordered polar structure caused by random fields. The maze adjacent and needle-like domains were visualized also by PFM (Fig. 2c,d). The maze adjacent domains were separated by 180° walls. The needle-like domains were more regular and narrow. It is worth noting that two kinds of domain walls with different orientations can also be obtained and are in contact at the interfaces close to the (110) planes. Generally, this plane is considered as the allowed orientation of the domains wall between certain ferroelectric-ferroelastic domain states. Thus, the needle-like shape corresponds to ferroelastic domains [14].

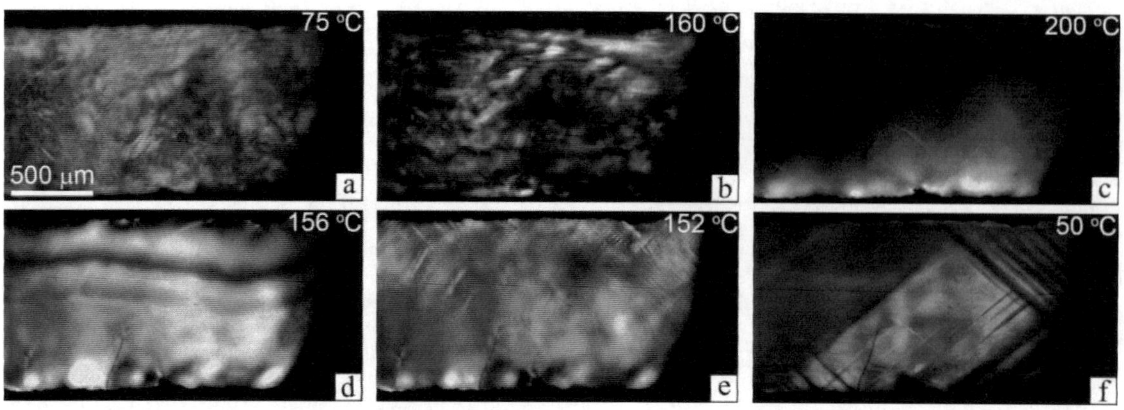

Figure 3. Main stages of domain structure evolution: (a), (b) zero field heating, (c) field annealing, (d), (e), (f) field cooling.

Figure 4. Sample after field treatment. PFM images: (a) topography, (b) amplitude, (c) phase. SEM images of (d) stripe and needle-like domains, (e) single domain regions.

3.2. The domain structure evolution during field treatment

The evolution of the domain structure in as-grown PMN-0.33PT sample was visualized *in situ* during consequent zero field heating, annealing in field, and field cooling by optical microscopy with crossed polarizers. The main stages of the process were defined (Fig. 3). First, destruction of as-grown microdomain structure in the temperature range of 75-90°C (Fig. 3a) and slow extinction of birefringent patterns up to 200°C (Fig. 3b) was obtained during zero field heating. Second, retention of the appeared transparent state was observed during field annealing at 200 °C (Fig. 3c). Third, slow appearance of birefringent patterns oriented in direction of the applied field was visualized during field cooling from 200 to 156°C (Fig. 3d). Fourth, the appearance and growth of needle-like domains was revealed during subsequent cooling from 156 to 150°C (Fig. 3e). Finally, the intensive domain growth from the sample edges (Fig. 3f) was observed. The observed non-180° in-plane domains balanced the inner strain during cooling. The amount of needle-like domains was quite small in the temperature range from 156 to 150 °C and cooling rate 0.1 °C/min, while the cooling rate increase for the temperatures below 120 °C led to fast increasing of the domain number. It is worth noting that highlighted stages do not fit perfectly to the dielectric permittivity maxima shown in Figure 1b. It could be explained by the shift of phase transition temperatures under the applied electric field [15].

3.3. Domain structure of poled sample

The sample after field treatment was polished with silica compound and the domain structure was visualized with PFM (Fig. 4a-c) and SEM (Fig. 4d-e). The domain structure after treatment became more uniform and regular. The stripe domains (width about 3 μm) and single domain regions (width about 20 μm) were formed. The needle-like domains became wider (width about 600 nm). The obtained increasing of the domain sizes as a result of field treatment allowed predicting achievement of the single domain state in (001) oriented tetragonal and (111) oriented rhombohedral PMN-PT crystals. Whereas, in studied 0.67PMN-0.33PT crystal the coexistence of *Ma* and *Mc* phases did not allow producing the desired single domain state.

4. Conclusion

The presented work demonstrated the effectiveness of application of complementary experimental methods of domain visualization for investigation of the domain structure evolution in multiaxial relaxor ferroelectric PMN-PT with high spatial resolution. The coexistence of maze adjacent 180° domains with antiparallel polarization and needle-like ferroelastic domains was revealed. Main stages of domain structure evolution during field treatment consisting of zero field heating, field annealing, and field cooling were defined. The transformation of as-grown domain structure to a more regular one as a result of field treatment was demonstrated.

Acknowledgments
The authors would like acknowledge Greshnyakov E. D. for sample preparation. The equipment of the Ural Center for Shared Use "Modern Nanotechnology" UrFU has been used. The research was made possible in part by Government of the Russian Federation (Act 211, Agreement 02.A03.21.0006) and RFBR (grant 17-52-80116-BRICS_a). V.S. acknowledges financial support within the State Task from the Ministry of Education and Science of the Russian Federation (Project No. 1366.2014/236).

References
[1] Park S E and Shrout T R 1997 *J. Appl. Phys.* **82** 1804-11
[2] Fu H and Cohen R E 2000 *Nature* **403** 281-3
[3] Zheng L, Lu X, Shang H, Xi Z, Wang R, Wang J, Zheng P and Cao W 2015 *Phys. Rev.* B **91** 184105
[4] Zhang S and Li F 2012 *J. Appl. Phys.* **111** 031301
[5] Zhang R, Jiang B and Cao W 2001 *J. Appl. Phys.* **90** 3471-5
[6] Zhang R, Jiang B and Cao W 2003 *IEEE Trans.Ultrason. Ferroelectr. Freq. Control.* **82** 3737-9
[7] Shur V Ya, Rumyantsev E L, Nikolaeva E V, Shishkin E I, Batchko R G, Fejer M M and Byer R L 2001 *Ferroelectrics* **257** 191-202
[8] Shur V Ya, Akhmatkhanov A R and Baturin I S 2015 *Appl. Phys. Rev.* **2** 040604
[9] Wang R, Xu H, Yang B, Luo Z, Sun E, Zhao J, Zheng L, Dong Y, Zhou H, Ren Y, Gao C and Cao W 2016 *Appl. Phys. Lett.* **108** 152905
[10] Wang L, Xu Z, Li Z and Li F 2010 *Mater. Sci. Eng.* B **170** 113-6
[11] Kuznetsov D K, Chezganov D S, Mingaliev E A, Kosobokov M S and Shur V Ya 2016 *Ferroelectrics* **503** 60-7
[12] Shur V Ya and Zelenovskiy P S 2014 *J. Appl. Phys.* **116** 066802
[13] Yu H, Zeng H, Chu R, Li G, Luo H, Xu Z and Yin Q 2004 *J. Phys. D: Appl. Phys.* **37** 2914-7
[14] Zeng H R, Yu H F, Chu R Q, Li G R, Luo H S and Yin Q R 2004 *J. Cryst. Growth* **267** 194-8
[15] Cao H, Bai F, Wang N, Li J and Viehland D 2005 *Phys. Rev.* B **72** 064104

IOP Conf. Series: Materials Science and Engineering **256** (2017) 012026 doi:10.1088/1757-899X/256/1/012026

Model for growth of fractal solid state surface and possibility of its verification by means of atomic force microscopy

D A Kulikov[1,*], A A Potapov[2,3], A E Rassadin[4] and A V Stepanov[5]

[1]P.G. Demidov Yaroslavl State University, 150003, Yaroslavl, Russia
[2]Kotel'nikov Institute of Radio Engineering and Electronics of RAS, 125009, Moscow, Russia
[3]Joint-Lab. of JNU-IREE RAS, JiNan University, China
[4]Nizhny Novgorod Mathematical Society, Nizhny Novgorod, 603950, Russia
[5]Chuvash State Agricultural Academy, 428003, Cheboksary, Russia

[*]brat_ras@list.ru

Abstract. In the paper, methods of verification of models for growth of solid state surface by means of atomic force microscopy are suggested. Simulation of growth of fractals with cylindrical generatrix on the solid state surface is presented. Our mathematical model of this process is based on generalization of the Kardar-Parisi-Zhang equation. Corner stones of this generalization are both conjecture of anisotropy of growth of the surface and approximation of small angles. The method of characteristics has been applied to solve the Kardar-Parisi-Zhang equation. Its solution should be considered up to the gradient catastrophe. The difficulty of nondifferentiability of fractal initial generatrix has been overcome by transition from a mathematical fractal to a physical one.

1. Introduction

Since the last quarter of the twentieth century, scanning probe microscopy (SPM) has broadened extensively because of arising of new research methods, such as piezoresponse force microscopy, apertureless scanning near-field optical microscopy, magnetic resonance force microscopy, and so on. Nowadays, the area of applications of SPM includes solid state physics, biology, mineralogy, etc. But since the progress in functional analysis, image processing, and computer technique is considerable, atomic force microscopy (AFM) invented by G. Binning, C.F. Quate, and C. Gerber in 1986 did not exhaust its possibilities yet, because using of AFM allows us to select adequate models for growing surface of the solid state.

Let us demonstrate this idea on the example of the Kardar-Parisi-Zhang equation (KPZ) [1]:

$$\frac{\partial z}{\partial t} = c \cdot \sqrt{1 + (\nabla z)^2} + \mu \cdot \nabla^2 z + Q(x, y, t) . \qquad (1)$$

The KPZ-equation has been applied to modelling of the process of sputtering of a substance on a surface of solid state [1]. In equation (1), z is the height of the surface – in general case, it is a function of coordinates $z(x, y, t)$; ∇ is two-dimensional gradient. The first term on the right-hand side of equation (1) shows that the growth of the surface occurs at a constant surface growing rate c strictly along the local normal to it. The second term on the right-hand side describes relaxation of the surface by a surface tension μ. At last, $Q(x, y, t)$ is an extra source of particles deposition.

Equation (1) should be provided by initial condition:

$$z(x,y,0) = z_0(x,y), \quad (x,y) \in D \subset R^2, \tag{2}$$

corresponding to initial shape of the surface under investigation.

It is clear that function (2) in domain D can be measured by means of AFM. After the finish of the sputtering process at the moment of time $t = T$ the height $\tilde{z}(x,y,T)$ of the surface in domain D can be estimated by AFM again. On the other hand, starting from initial condition (2), one can solve the KPZ-equation (1) numerically on the interval of time $t \in [0,T]$ and derive the theoretical height $z(x,y,T)$ of the surface at $t = T$. Therefore, we can compare these two heights in some norm, for instance in the functional space $C(\overline{D})$ of functions continuous on $\overline{D}$:

$$\| z - \tilde{z} \| = \sup_{(x,y)\in D} | z(x,y,T) - \tilde{z}(x,y,T) |. \tag{3}$$

If number (3) is greater than some earlier chosen threshold value, then one claims that the theoretical model of growth is wrong; hence, it should be rejected or modified. Thus, in order to use this method of comparison of the theory and experiment, it is required to develop various generalizations of the KPZ-equation (1).

For simplicity in this paper, we confine ourselves to an analysis of the growth of a surface with a cylindrical generatrix (Fig. 1).

It is necessary to note here that fractal phenomena are sure to be widespread in physics of solid state. In particular, in work [2] based on the study of the surface images obtained with AFM for a number of samples of structural materials subjected to typical physical-chemical processing, such as microarc oxidation, diamond grinding etc., it is uniquely shown that the investigated surfaces are fractal. Effects of fractality on surface of nanostructured semiconductors have been described in [3]. Formation of fractal patterns on the surface of nanocomposite materials is under investigation in [4]. In [5], it is proved that in metals the surface layer has a fractal structure. It is obvious that there are some features under solution of the KPZ-equation (1) with fractal initial condition (2) in comparison with the same one with smooth initial condition (2).

The rest of the article is organized as follows: in Section 2 we demonstrate how to construct the generalized KPZ-equation in approximation of small angles taking into account the assumption about the anisotropy of the surface growth [6]. Section 3 deals with the regularized Darboux function. Peculiarities of mathematical modeling of the growth of the fractal surface in the framework of the method of characteristics are under consideration in Section 4. Perspectives of further investigations are also discussed.

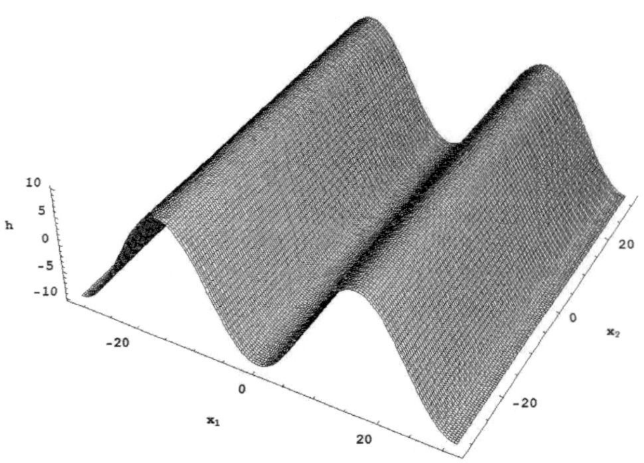

Figure 1. A surface with a cylindrical generatrix.

2. Derivation of generalized KPZ-equation in the framework of approximation of small angles

The aim of this section is to obtain a nonlinear partial differential equation of the first order for the height of anisotropically growing surface. In this case, the generalized KPZ-equation has the form [7]:

$$\frac{\partial z}{\partial t} = V(\theta)\cdot\left[1+\left(\frac{\partial z}{\partial x}\right)^2\right]^{\frac{1}{2}}, \quad z(x,0)=z_0(x), \qquad x\in R, \tag{4}$$

where $z(x,t)$ – the height of growing surface with a cylindrical generatrix, $V(\theta)$ – the rate of its growth, which depends on the angle θ between the local normal to the surface and the z-axis, and $z_0(x)$ is initial shape of the surface under investigation. Surface tension and extra source are not considered in this model.

Wherein

$$V(\theta)=\int_{\theta_-(\theta)}^{\theta_+(\theta)} D(\theta')\cdot\cos(\theta-\theta')\cdot d\theta', \tag{5}$$

where $D(\theta)$ – a directivity diagram that determines the intensity of the flow of particles falling out on the growing surface [7], functions $\theta_\pm(\theta)$ take into account the shading effects of the rest of the profile $z(x,t)$, and angle θ is related to the profile under consideration by the next relation [7]:

$$\tan\theta = -\frac{\partial z}{\partial x}. \tag{6}$$

The anisotropy of the growth of the surface is considered by means of the following expression for the directivity diagram:

$$D(\theta)=D_0+D_2\cdot\cos^2\theta, \; D_0>0, \; D_2>0. \tag{7}$$

A typical graph of such directivity diagram is presented in Figure 2.

At first, let us calculate the rate (5) of surface growth in our case. Suppose that there is no shading by further parts of generatrix $z(x,t)$, then the limits of integration in formula (5) are equal to:

$$\theta_-(\theta)=\begin{cases}\theta-\pi/2, & \theta>0\\ -\pi/2, & \theta<0\end{cases} \qquad \theta_+(\theta)=\begin{cases}\pi/2, & \theta>0\\ \pi/2+\theta, & \theta<0\end{cases}. \tag{8}$$

Combining expressions (5), (7), and (8) we find that:

$$V(\theta)=D_0\cdot(1+\cos\theta)+\frac{D_2}{3}\cdot(1+\cos\theta)^2. \tag{9}$$

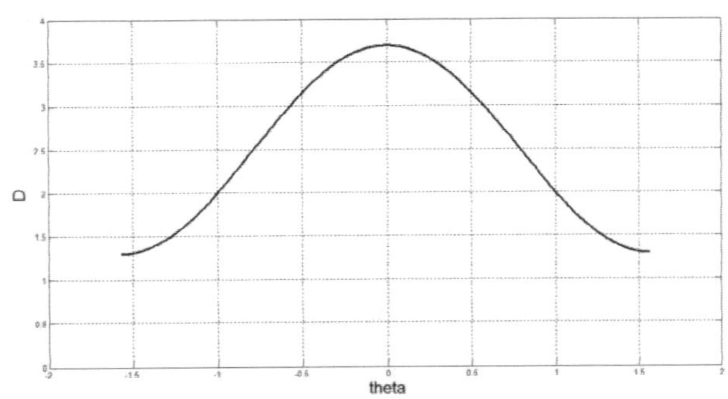

Figure 2. Angular dependence of directivity diagram.

Secondly, applying connection (6) between angle θ and generatrix $z(x,t)$ one can rewrite equation (4) as follows:

$$\frac{\partial z}{\partial t} = D_0 + \frac{2\cdot D_2}{3} + \left(D_0 + \frac{D_2}{3}\right)\cdot\left[1+\left(\frac{\partial z}{\partial x}\right)^2\right]^{\frac{1}{2}} + \frac{D_2}{3}\cdot\left[1+\left(\frac{\partial z}{\partial x}\right)^2\right]^{-\frac{1}{2}}. \tag{10}$$

After that let us assume that $|\theta| << 1$. This is the so-called approximation of small angles. In the framework of this approximation $|\partial z/\partial x| << 1$. Hence, using Taylor expansion of the right side of equation (10) and rejecting terms with powers of $\partial z/\partial x$ greater than four we can simplify this equation in the next manner:

$$\frac{\partial z}{\partial t} = 2\cdot D_0 + \frac{4\cdot D_2}{3} + \frac{D_0}{2}\cdot\left(\frac{\partial z}{\partial x}\right)^2 + \frac{2\cdot D_2 - 3\cdot D_0}{24}\cdot\left(\frac{\partial z}{\partial x}\right)^4. \tag{11}$$

One can eliminate homogeneous growth of the surface from equation (11) by trivial substitution:

$$z(x,t) = \left(2\cdot D_0 + \frac{4\cdot D_2}{3}\right)\cdot t + h(x,t), \tag{12}$$

which leads to the following Cauchy problem for new unknown function $h(x,t)$:

$$\frac{\partial h}{\partial t} = \frac{D_0}{2}\cdot\left(\frac{\partial h}{\partial x}\right)^2 + \frac{2\cdot D_2 - 3\cdot D_0}{24}\cdot\left(\frac{\partial h}{\partial x}\right)^4, \quad h(x,0) = z_0(x). \tag{13}$$

We underline that in order to get contributions of the same order from both the first and the second terms in the right side of equation (13) the inequality $D_2 >> D_0$ ought to be true.

Let us choose for initial condition for equation (13) typical values of its amplitude A and spatial scale L namely:

$$z_0(x) = A\cdot h_0(x/L), \tag{14}$$

where $h_0(\ldots)$ is dimensionless function then rescaling time t, coordinate x, and nonhomogeneous height of growing surface h as $\frac{A\cdot D_0\cdot t}{L^2} \to t$, $\frac{x}{L} \to x$ and $\frac{h}{A} \to h$ we reduce equation (13) and its initial condition (11) to the dimensionless form:

$$\frac{\partial h}{\partial t} = \frac{1}{2}\cdot\left(\frac{\partial h}{\partial x}\right)^2 + \frac{g}{4}\cdot\left(\frac{\partial h}{\partial x}\right)^4, \quad h(x,0) = h_0(x), \tag{15}$$

where $g = \frac{2\cdot D_2 - 3\cdot D_0}{6\cdot D_0}\cdot\frac{A^2}{L^2}$ is a positive coupling constant.

The solution of Cauchy problem (15) by the method of characteristics (see [7] and references therein) is equal to:

$$x = y + (1 + g\cdot u_0^2(y))\cdot u_0(y)\cdot t \quad h = h_0(y) - \left(\frac{1}{2} + \frac{3}{2}\cdot g\cdot u_0^2(y)\right)\cdot u_0^2(y)\cdot t, \tag{16}$$

where $u_0(y) = -h_0'(y)$.

3. The regularized Darboux function as an example of physical fractal

It is well-known that fractal functions are continuous nowhere differentiable ones (see [8] and references therein). The formulas (16) for solution of nonlinear partial differential equation (15) include not only the initial profile $h_0(x)$, but also its derivative $h_0'(x)$. This means that to describe the growth of fractal surfaces within the framework of the approach developed in the previous Section, the fractal initial profile must undergo some regularization.

The conceptual basis of this regularization is known to include in transfer from mathematical fractals to physical fractals [8]. The scheme of this transfer is presented in Figure 3. We stress that this scheme was approved by B. Mandelbrot [9].

In order to illustrate this approach by concrete example, let us consider the Darboux function [8]:

$$D(x) = \sum_{n=1}^{\infty} \frac{\sin[(n+1)! \cdot x]}{n!}. \quad (17)$$

This function is nondifferentiable for all real numbers [8]; however, if we take only a limited number N of terms of the series (17), then the resulting function:

$$D_N(x) = \sum_{n=1}^{N} \frac{\sin[(n+1)! \cdot x]}{n!} \quad (18)$$

will be differentiable as many times as required.

Error ε_N of this approximation:

$$| D(x) - D_N(x) | \le \varepsilon_N, \quad (19)$$

can be easily estimated with help of inequality $|\sin x| \le 1$ which is valid for all real x namely

$$\varepsilon_N = \sum_{n=N+1}^{\infty} \frac{1}{n!} \quad (20)$$

For sum (20) from any course of calculus the following inequality is well known:

$$\varepsilon_N \le \frac{1}{N! \cdot N} \quad (21)$$

Thus, from the formula (21) one can see that under quite large N error of approximation of the Darboux function (17) by means of truncated functional series (18) becomes quite small.

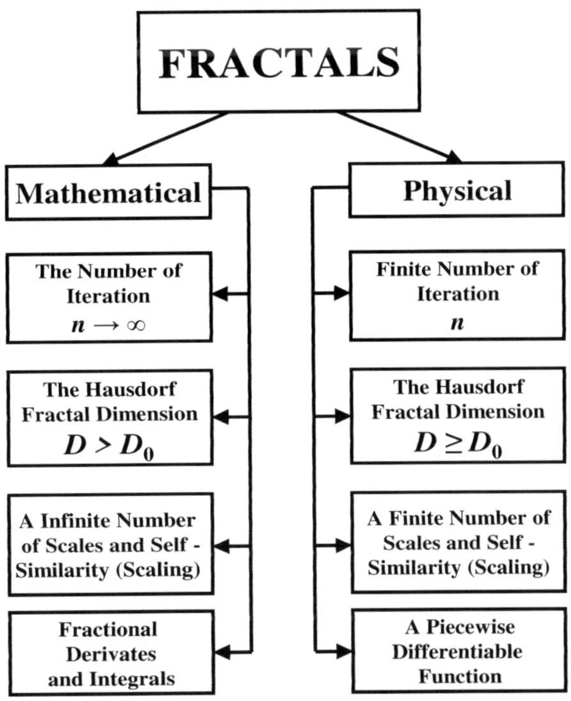

Figure 3. The Potapov scheme of mathematical and physical fractals.

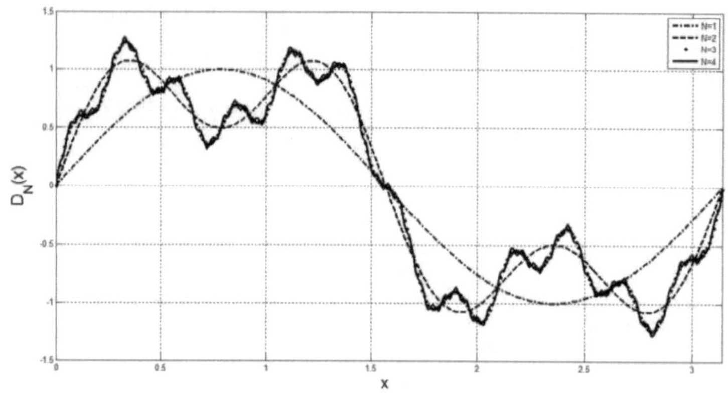

Figure 4. The regularized Darboux functions for N = 1,2,3,4.

Graphs of the regularized Darboux functions (18) for N = 1,2,3,4 are shown in Figure 4. One can observe that there are no sharp distinctions between $D_3(x)$ and $D_4(x)$ yet.

The regularized Darboux function possesses by the following derivative:

$$D'_N(x) = \sum_{n=1}^{N} (n+1) \cdot \cos[(n+1)! \cdot x] . \tag{22}$$

It is obvious that there is the next inequality for upper bound of function (22):

$$| D'_N(x) | \le \frac{N \cdot (N+3)}{2} , \tag{23}$$

Therefore, absolute value of the derivative of the regularized Darboux function grows as $O(N^2)$ when N tends to infinity. This circumstance seems to look like violation of correctness of approximation of small angles, but we emphasize that there is no contradiction indeed, because requirement $| \partial h / \partial x | << 1$ was imposed on dimensionfull variables in dimensionfull equation (13). It means that in accordance with inequality (23) for self-consistency of our assumptions typical values of amplitude A and spatial scale L of dimensionfull initial condition (14) ought to obey to the following inequality:

$$\frac{A \cdot N \cdot (N+3)}{2 \cdot L} << 1 . \tag{24}$$

4. Results of numerical simulation

Let us now demonstrate the results of numerical modeling for the above developed approaches.

As a dimensionless initial condition, the regularized Darboux function (18) under N=4 was chosen:

$$h_0(x) = D_4(x) . \tag{25}$$

In this case, the error of approximation of mathematical fractal (17) by physical fractal (25) is equal or less than $1/96$: $\varepsilon_4 \le 1/96 \approx 0.0104$. This level of error is expected to be appropriate for rigor of our consideration. Further, due to periodicity of initial condition (25) namely $D_4(x+\pi) = D_4(x)$ we can restrict ourselves by the investigation of the behavior of the surface over the segment $[0, \pi]$ of the x-axis without loss of generality. Value of coupling constant is equal to g=0.01.

The initial condition (25) and the result of its temporal evolution at t=0.0005 are presented in Figure 5 above and below, respectively. From this Figure, one can see that temporal evolution leads to narrowing of the profile of the surface in the neighborhood of local maxima of the function $h(x,t)$.

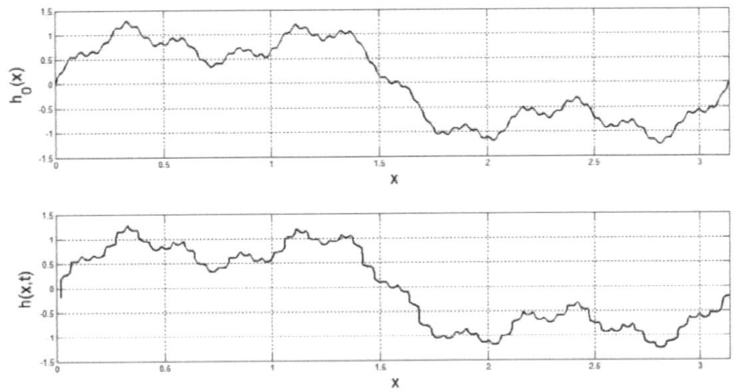

Figure 5. Initial height of cylindrical generatrix (above) and result of its temporal evolution until gradient catastrophe (below).

We underline that researchers ought to be neat and careful under operating with equation (15), because representation (16) of solution of this equation is applicable only up to the moment of gradient catastrophe. At this moment of time, one-to-one correspondence $y(x,t)$ between Eulerian coordinate x and Lagrangian coordinate y in accordance with the first formula in formula (16) is violated. In order to determine moment of gradient catastrophe for fixed initial profile, it is necessary to watch zeroes of the following Jacobian:

$$J(y,t) \equiv \frac{\partial x(y,t)}{\partial y} = 1 + (1 + 3 \cdot g \cdot u_0^2(y)) \cdot u_0'(y) \cdot t \,. \tag{26}$$

This untrivial situation is illustrated by Figure 6. On the left side of Figure 6 $J(y,t)$ and $x(y,t)$ at t=0.0005 corresponding to Figure 5 are presented. In this case, Jacobian (26) is positive everywhere $J(y,t) > 0$ and, therefore, function $y(x,t)$ is one-to-one mapping. On the right side of Figure 6, $J(y,t)$ and $x(y,t)$ at t = 0.006 are presented. In this case, one can observe that there are a lot of zeroes of Jacobian (26) and, hence, $y(x,t)$ is multifunction, evidently. It means that at t = 0.006 instead of equations (16) one ought to deal with the so-called weak solutions of equation (15) constructed in the framework of the Oleynik-Lax absolute minimum principle or the E-Rykov-Sinai global principle (see [7] and references therein).

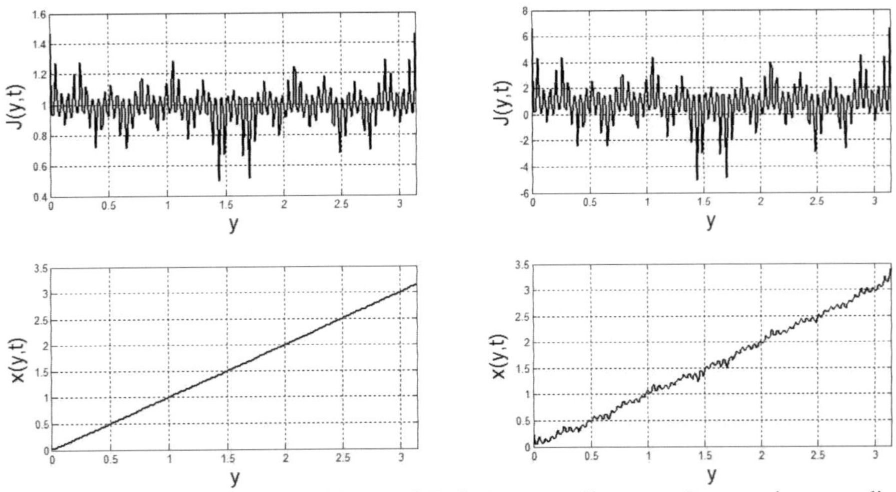

Figure 6. Dependences of Jacobian and Eulerian coordinate on Lagrangian coordinate before (on the left side) and after (on the right side) of gradient catastrophe.

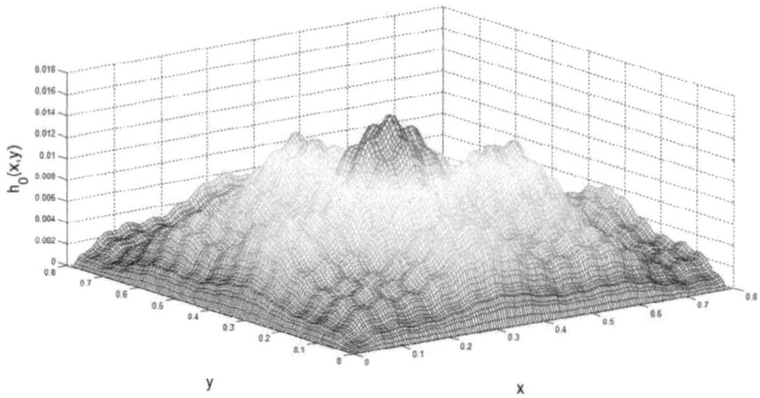

Figure 7. The two-dimensional regularized Darboux function.

The aim of further investigation is thought to extend the approach elaborated above on general two-dimensional situation both before and after gradient catastrophe. For instance, in two-dimensional case the initial profile (18) can be generalized as follows:

$$h_0(x, y) = D_N(x) \cdot D_N(y). \tag{27}$$

Physical fractal (27) under N=5 is shown in Figure 7.

After that, fractal dimension of growing surface ought to be calculated in accordance with solution of the two-dimensional generalized KPZ-equation by methods developed earlier [8]. One can compare the obtained result with estimation of fractal dimension based on data of AFM for the surface under investigation. Another way of comparison of the theory and AFM-measurements is to investigate connections between fractal properties of the surface of solid state and scattering data on this surface of electromagnetic and elastic waves [10]. Beyond any doubt both of these criteria ought to be analyzed simultaneously with norm (3). In order to broaden sphere of applications of the suggested verification procedure, we also intend to consider other examples of equations describing the growth of surface of solid state, for instance the Bradley-Harper equation [11], which is a nonlinear mathematical model for the formation of an inhomogeneous topography on the surface of solid state under ionic bombardment.

5. Conclusion

In the article, procedure of verifying models of growth of solid state surface and selecting the true ones with support of this process by AFM has been considered. To realize the procedure in practice, the exact solution of the generalized KPZ-equation describing anisotropic growth of the surface with cylindrical generatrix until the gradient catastrophe has been constructed in the framework of approximation of small angles. In order to demonstrate efficiency of the suggested procedure, the regularized Darboux function has been chosen as the initial generatrix.

Acknowledgements
The authors prepared this article during working on the project "Leading Talents of Guangdong Province", No. 00201502 (2016-2020) in the JiNan University (China, Guangzhou).

References

[1] Kardar M, Parisi G and Zhang Y C 1986 *Phys. Rev. Lett.* **56** 889
[2] Potapov A A, Bulavkin V V, German V A and Vyacheslavova O F 2005 *Techn. Phys.* **50** 560
[3] Zhanabaev Z Zh and Grevtseva T Yu 2014 *Rev. of Theor. Sci.* **2** 211
[4] Bahmatskaya A I and Plugotarenko N K 2014 *Izvestiya SFedU. Eng. Sci.* **9** 118 (in Russian).
[5] Oleshko V S and Yurov V M 2017 *Eurasian Phys. Techn. J.* **14** 75

[6] Barabasi A L and Stanley H E 1995 *Fractal Concepts in Surface Growth* (Cambridge: Cambridge University Press)
[7] Gurbatov S N, Rudenko O V and Saichev A I 2011 *Waves and Structures in Nonlinear Nondispersive Media: General Theory and Applications to Nonlinear Acoustics (Nonlinear Physical Science)* (Springer)
[8] Potapov A A, Gulyaev Yu V, Nikitov S A, Pakhomov A A and German V A 2008 *The Modern Methods of Image Processing* ed A.A. Potapov (Moscow: FIZMATLIT) (in Russian)
[9] Potapov A A 2007 *Nonlinear World* **5** 402 (in Russian)
[10] Potapov A A 2017 On the Indicatrixes of Waves Scattering from the Random Fractal Anisotropic Surface *Fractal Analysis - Applications in Physics, Engineering and Technology* ed F. Brambila (Rijeka: InTech) chapter 9 pp 187-248
[11] Bradley R M and Harper J M E 1988 *J. Vac. Tech.* A **6** 2390

IOP Conf. Series: Materials Science and Engineering **256** (2017) 012027 doi:10.1088/1757-899X/256/1/012027

The phase-field modeling of the self-organized phase growth with three-fold symmetry

A R Akhmatkhanov[*], A I Lobov, M A Chuvakova, E D Saveliev and V Ya Shur

School of Natural Sciences and Mathematics, Ural Federal University, 620000 Ekaterinburg, Russia

[*]andrey.akhmatkhanov@urfu.ru

Abstract. The formation of self-organized domain patterns during polarization reversal in highly non-equilibrium switching conditions was studied by us earlier in the single crystals of lithium niobate and lithium tantalate. In this paper, we used the phase-field simulation to verify the analogy between self-organized growth of domains and new phase during the first order phase transition. The crystal symmetry C_{3v} was taken into account. The similarity of simulated and experimentally observed shapes of isolated domains was achieved.

1. Introduction

The phase-field model is widely used as a powerful computational method for modeling and predicting morphological and microstructure evolution in materials [1]. It has been shown that the model can be used successfully for studying of the formation of self-assembled structures during the first order phase transitions in highly non-equilibrium conditions, for example for description of dendritic crystal growth [2, 3]. The formation of self-assembled structures including the dendrite ones was demonstrated experimentally [4-8] and the phase diagram of the most common structure types was predicted [2,9,10]. However, the first order phase transitions including the crystallization from the melt possess the limited range of control parameters thus prohibiting the realization of all predicted types of structure in one system. This fact stimulates searching new systems with simple variation of control parameters. It is known that the evolution of ferroelectric domains during polarization reversal can be considered as an analog of the first-order phase transformation with electric field as a driving force [11-14]. The highly non-equilibrium switching conditions can be realized in ferroelectrics in wide range of control parameters. The complementary methods of domain visualization with high spatial and temporal resolutions are available [15-17]. The growth of self-organized dendrite domain patterns in ferroelectric lithium niobate ($LiNbO_3$, LN) and lithium tantalate single crystals with C_{3v} symmetry was demonstrated by us earlier [4-8, 18]. Their great practical importance enabled the growth of the large crystals with extremely high uniformity. It was shown that the formed structures could demonstrate the C_3 and the C_6 envelope symmetry depending on the growth conditions [4-8, 18].

In this paper, we present the results of application of phase-field model to the domain growth to verify the analogy between self-organized growth of domains and new phase during the first order phase transition. The C_{3v} symmetry of the crystals was taken into account.

2. Experimental results

Two self-organized domain patterns appeared during polarization reversal in LN single crystals have been analyzed.

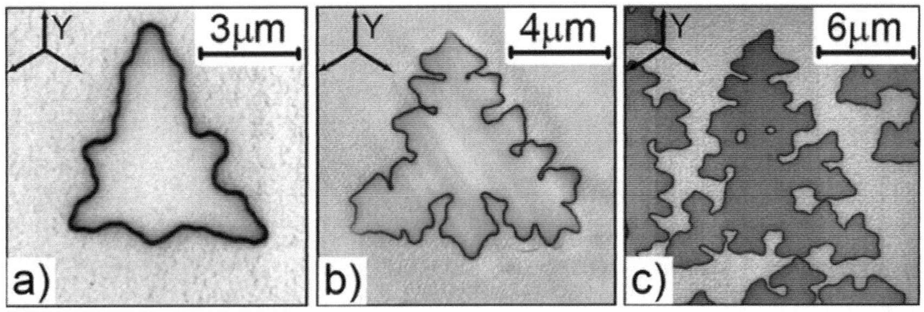

Figure 1. Piezoelectric force microscopy (PFM) image of dendrite structures obtained in stoichiometric LN at 230^{0}C for various numbers of rectangular field pulses: (a) single pulse E_{max}=1.6 kV/mm, (b) two pulses E_{max}=1.6 kV/mm, (c) three pulses E_{max}=1.5 kV/mm. Reprinted with permission from [5]. Copyright 2012, AIP Publishing LLC.

The first object was the dendrite domains appeared in stoichiometric LN crystals at the temperatures above 230°C (Fig. 1) [5]. After partial switching from single domain state, the dendrite domains with sizes ranging from 3 to 30 μm were distributed over the Z polar surface covered by the electrode. The domain shape was close to six-ray stars with rays along crystallographic Y axes (Fig. 1). Y+ oriented rays were essentially longer than Y- ones. Several field pulses led to formation of more complicated dendrite structures (Fig. 1b,c).The second object was the domains growing in LN with surface layer modified by proton exchange (PE). They formed the self-organized sub-micron scale dendrite domain patterns consisting of the stripes oriented along the X crystallographic directions separated by the arrays of dashed residual domains (Fig. 2) [18]. The domain stripes formed as a result of branching of three Y oriented rays. The branching period was about 2 μm, close to the thickness of the PE layer. This process led to formation of the self-organized stripe structure with period of the dashed structure from 2 to 3 μm.

The obtained results demonstrate that the envelopes of self-organized domain structures in LN can follow both the crystal cymmetry C_3 (Fig. 1) and the higher symmetry C_6 (Fig. 2).

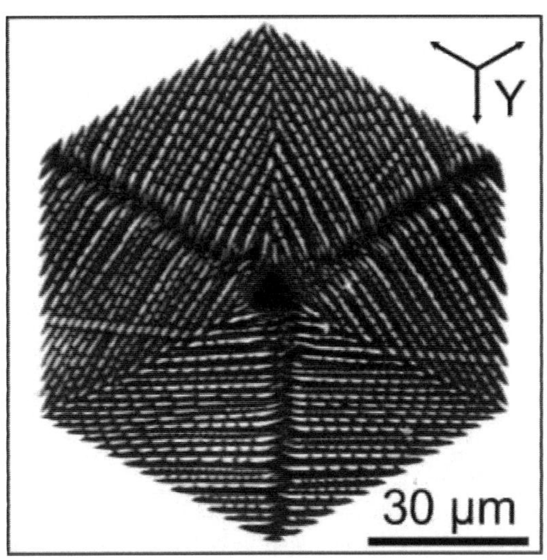

Figure 2. The self-organized domain structure obtained at the Z+ surface of LN with Z- surface modified by PE. PFM, phase signal. Dark area corresponds to the switched domains. Reprinted with permission from [18]. Copyright 2017, AIP Publishing LLC.

IOP Conf. Series: Materials Science and Engineering **256** (2017) 012027 doi:10.1088/1757-899X/256/1/012027

3. Phase-field modeling

The formation of the self-organized domain patterns was analyzed within the phase-field model for one component crystallization from the melt based on the analogy between the sideways motion of domain walls and the phase boundary motion during the first order phase transformations [19]. In this case, the main factor limiting the phase boundary motion velocity was the local temperature increase due to release of crystallization latent heat.

We have used the exothermal model of the first order phase transformation based on the time-dependent Ginzburg-Landau equation and the heat conductivity equation [2]. The model included two variables: the phase-field $p(r,t)$ and the temperature $T(\mathbf{r},t)$. The value $p(\mathbf{r},t)=0$ corresponds to the liquid phase and the $p(\mathbf{r},t)=1$ – to the solid phase. The phase boundary corresponds to a narrow layer, where $p(\mathbf{r},t)$ values lie between 0 and 1.

The following expression of Ginzburg-Landau free energy have been used:

$$\Phi(p,m) = \int (\tfrac{1}{2}\, \varepsilon^2 |\nabla p|^2 + F(p,m))d\mathbf{r} \tag{1}$$

where ε is a small parameter determining the width of phase boundary, m is parameter determining the oversaturation, F is two-well potential with local minima at $p = 0$ and $p = 1$ for any m. For the phase-field models of such type, the F function is usually chosen as [2]:

$$F(p,m) = \tfrac{1}{4}\, p^4 - (\tfrac{1}{2} - m/3)p^3 + (\tfrac{1}{4} - \tfrac{1}{2}\, m)\, p^2 \tag{2}$$

Based on the kinetic equation $\tau\partial p/\partial t = -\delta\Phi/\delta p$, we will get for the 2D problem taking into account that ε parameter is a function of angle between the normal to phase boundary and certain predefined direction ($\varepsilon=\varepsilon(\theta)$):

$$\tau\frac{\partial p}{\partial t} = -\frac{\partial}{\partial x}\left(\varepsilon\varepsilon'\frac{\partial p}{\partial y}\right) + \frac{\partial}{\partial y}\left(\varepsilon\varepsilon'\frac{\partial p}{\partial x}\right) + \nabla(\varepsilon^2\nabla p) + p(1-p)(p - \frac{1}{2} + m) \tag{3}$$

where τ is a small positive constant and the prime means the derivative $d/d\theta$. According to the obtained experimental results, the anisotropy of the parameter $\varepsilon(\theta)$ was chosen as a sum of C_3 and C_6 symmetry terms:

$$\varepsilon(\theta) = \varepsilon_0\, (1+\delta(\beta\cos(3\theta)+(1-\beta)\cos(6\theta))) \tag{4}$$

where ε_0 is an average value of interface energy, δ is anisotropy value, and β is a relative input of C_3 symmetry term.

The following formula was used for temperature dependence of oversaturation:

$$m(T) = (\alpha/\pi)\arctan(\gamma(T_e-T)) \tag{5}$$

where α and γ are the positive constants ($\alpha<1$), T_e is an equilibrium temperature, T is a local temperature.

The temperature in each point was calculated using the heat conductivity equation at the boundary with additional heat source:

$$\frac{\partial T}{\partial t} = \nabla^2 T + K\frac{\partial p}{\partial t} \tag{6}$$

where K is normalized latent heat of crystallization.

The computer modeling consisted in the solving of the system of differential equations (3) and (6). For the initial moment, the phase-field $p(\mathbf{r},0)$ corresponded to one small circular region of solid phase in the center and the temperature in the whole volume was equal to T_e.

We have studied also the influence of noise on the obtained structure parameters. The additional term $ap(1-p)\chi$, where a is noise amplitude, χ – random number uniformly distributed in the range [-½;½], was added to the equation (3).

The parameter values, used for the computer modeling are presented in Table 1.

IOP Conf. Series: Materials Science and Engineering **256** (2017) 012027 doi:10.1088/1757-899X/256/1/012027

The obtained self-organized structures are presented in Figure 3. It is clearly seen that the increase of relative input of C_3 term leads to transition from the hexagonal envelope shape to the triangular one.

We have analyzed the obtained transition in terms of change of the envelope shape constant C (Fig. 4) calculated as:

$$C = S/R^2 \tag{7}$$

where S is the area of obtained envelope, R is the envelope radius corresponding to the maximum distance between the center of the structure and its point at the border.

Table 1. The values of the parameters used for the computer simulation.

Parameter	Value
Anisotropy terms	C_3 and C_6
Input of C_3 symmetry (β)	from 0 to 1
Latent heat (K)	1.5
Equilibrium temperature (T_{eq})	0.9
Noise amplitude (a)	0.01
Average interface energy (ε_0)	0.1
Anisotropy value (δ)	0.2
The constants for calculating the oversaturation (α and γ)	0.9 and 10
τ parameter	0.0003

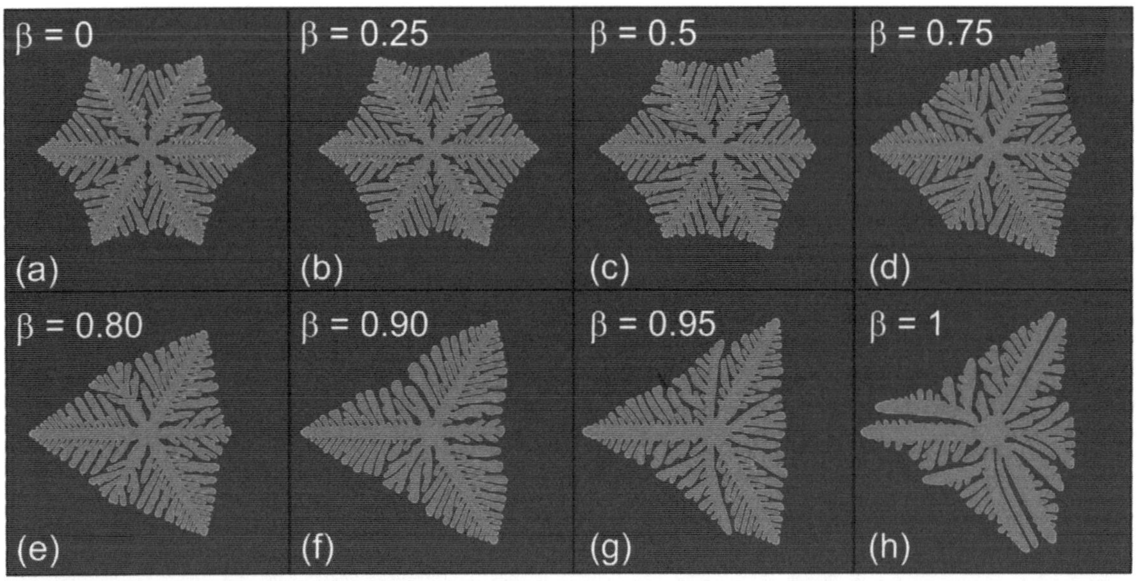

Figure 3. (a)-(h) The self-organized structures obtained as a result of phase-field computer simulation with parameters presented in Table 1. Only β parameter was changed during the subsequent simulations.

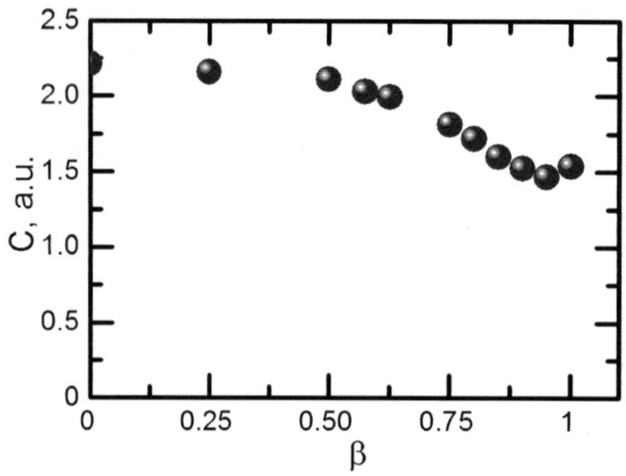

Figure 4. The dependence of the envelope shape constant calculated by equation (7) on the relative input of the C_3 symmetry term β.

The envelope shape constant decreases from the value about 2.3 for hexagonal shape to the value about 1.5 for the triangular shape (Fig. 4). The slight increase of the shape constant near $\beta = 1$ can be related to the formation of doublons oriented along C_3 symmetry axes (Fig. 3h).

The influence of the noise on the symmetry of obtained structures has been studied. The structures with parameters presented in Table 1 and the pure C_6 symmetry ($\beta = 0$, see Figure 3a) and the noise amplitude a from 0 to 0.012 have been analyzed. The symmetry of the structure was estimated according to the following procedure. The binary image of the structure was rotated to the angle 60° with respect to the center of the image. The overlap integral between the initial image and the rotated one was calculated. The overlap integral was normalized to the whole area of the structure. Obtained dependence of normalized overlap integral (I) on the noise value is presented in Figure 5. The increase of the noise leads to decrease of the structure symmetry. The relatively high values of the normalized overlap integral for all studied noise amplitudes demonstrates system's stability against noise.

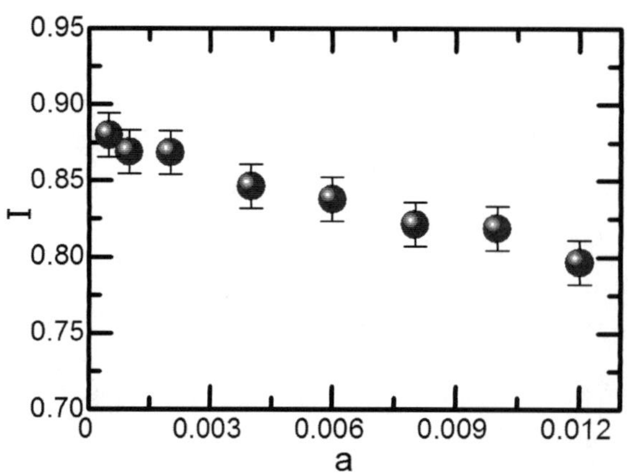

Figure 5. The dependence of normalized overlap integral on the noise amplitude.

4. Conclusion

The self-organized phase growth with three-fold symmetry has been studied by computer simulation based on the phase-field model. The similarity of simulated and experimentally observed shapes of isolated domains has been achieved. It has been shown that by varying the relation between the C_3 and C_6 symmetry terms in the interface energy anisotropy formula we can obtain the C_3 and C_6 envelope symmetries of the obtained structures. It has been shown that the increase of the noise value leads to the decrease of the structure symmetry. However, the relatively high values of the normalized overlap integral for all studied noise amplitudes demonstrate system's stability against noise.

Acknowledgements

The equipment of the Ural Center for Shared Use "Modern nanotechnology" UrFU was used. The research was made possible by Russian Science Foundation (Project №14-12-00826).

References

[1] Chen L-Q 2002 Phase-field models for microstructure evolution *Annu. Rev. Mater. Res.* **32** 113-40
[2] Kobayashi R 1993 Modeling and numerical simulations of dendritic crystal growth *Phys. D Nonlinear Phenom.* **63** 410-23
[3] Wheeler A A, Murray B T and Schaefer R 1993 Computation of dendrites using a phase field model *Phys. D Nonlinear Phenom.* **66** 243-62
[4] Shur V Ya, Nikolaeva E V, Shishkin E I, Chernykh A P, Terabe K, Kitamura K, Ito H and Nakamura K 2002 Domain shape in congruent and stoichiometric lithium tantalate *Ferroelectrics* **269** 195-200
[5] Shur V Ya, Chezganov D S, Nebogatikov M S, Baturin I S and Neradovskiy M M 2012 Formation of dendrite domain structures in stoichiometric lithium niobate at elevated temperatures *J. Appl. Phys.* **112** 104113
[6] Shur V Ya and Akhmatkhanov A R 2017 Domain shape instabilities and dendrite domain growth in uniaxial ferroelectrics *Phil.Trans.R. Soc.* A *in press* (doi: 10.1098/rsta.2017.0204)
[7] Shur V Ya, Akhmatkhanov A R and Pelegova E V 2016 Self-organizing formation of dendrite domain structures in lithium niobate and lithium tantalate crystals *Ferroelectrics* **500** 76-89
[8] Shur V Ya, Kosobokov M S, Mingaliev E A, Kuznetsov D K and Zelenovskiy P S 2016 Formation of snowflake domains during fast cooling of lithium tantalate crystals *J. Appl. Phys.* **119** 144101
[9] Brener E 2000 Structure formation in diffusional growth and dewetting *Solid State Ionics* **131** 23-33
[10] Brener E, Müller-Krumbhaar H and Temkin D 1996 Structure formation and the morphology diagram of possible structures in two-dimensional diffusional growth *Phys. Rev.* E **54** 2714-22
[11] Shur V Ya, Gruverman A L and Rumyantsev E L 1990, Dynamics of domain structure in uniaxial ferroelectrics *Ferroelectrics* **111** 123-31
[12] Shur V Ya 1996 Fast polarization reversal process: evolution of ferroelectric domain structure in thin films *Ferroelectric Thin Films: Synthesis and Basic Properties. Ferroelectricity and Related Phenomena, vol. 10* ed C A Paz de Araujo, J F Scott and G W Taylor (Amsterdam: Gordon & Breach Science Publ.) pp 153-192
[13] Shur V Ya 2005 Correlated nucleation and self-organized kinetics of ferroelectric domains *Nucleation Theory and Applications* Ed J W P Schmelzer (WILEY-VCH, Weinheim) pp 178-214
[14] McGilly L J, Yudin P, Feigl L, Tagantsev A K and Setter N 2015 Controlling domain wall motion in ferroelectric thin films *Nat. Nanotechnol.* **10** 145-50
[15] Zelenovskiy P S, Shur V Ya, Kuznetsov D K, Mingaliev E A, Fontana M and Bourson P 2011 Visualization of nanodomains in lithium niobate single crystals by scanning laser confocal

IOP Conf. Series: Materials Science and Engineering **256** (2017) 012027 doi:10.1088/1757-899X/256/1/012027

Raman microscopy *Phys. Solid State* **53** 109-13

[16] Soergel E 2005 Visualization of ferroelectric domains in bulk single crystals *Appl. Phys.* B **81** 729-51

[17] Shur V Ya and Zelenovskiy P S 2014 Micro- and nanodomain imaging in uniaxial ferroelectrics: joint application of optical, confocal Raman and piezoelectric force microscopy *J. Appl. Phys.* **116** 066802

[18] Shur V Ya, Akhmatkhanov A R, Chuvakova M A, Dolbilov M A, Zelenovskiy P S and Lobov A I 2017 Formation of self-organized domain structures with charged domain walls in lithium niobate with surface layer modified by proton exchange *J. Appl. Phys.* **121** 104101

[19] Shur V Ya and Rumyantsev E L 1993 Crystal growth and domain structure evolution *Ferroelectrics* **142** 1-7